Monika Felchner-Żwirełło

Propionic Acid Degradation by Syntrophic Bacteria During Anaerobic Biowaste Digestion

Propionic Acid Degradation by Syntrophic
Bacteria During Anaerobic Biowaste Digestion

Karlsruher Berichte zur Ingenieurbiologie
Band 49

Institut für Ingenieurbiologie und Biotechnologie des Abwassers
Karlsruher Institut für Technologie

Herausgeber: Prof. Dr. rer. nat. J. Winter

Propionic Acid Degradation by Syntrophic Bacteria During Anaerobic Biowaste Digestion

by
Monika Felchner-Żwirełło

Dissertation, Karlsruher Institut für Technologie (KIT)
Fakultät für Fakultät für Bauingenieur-, Geo- und Umweltwissenschaften
Tag der mündlichen Prüfung: 08. Februar 2013
Referenten: Prof. Dr. rer. nat. habil. Josef Winter
Korreferenten: Prof. Dr.-Ing. E.h. Hermann H. Hahn, Ph.D.
                         Prof. Dr hab. inż. Jacek Namieśnik

Impressum

Scientific
Publishing

Karlsruher Institut für Technologie (KIT)
KIT Scientific Publishing
Straße am Forum 2
D-76131 Karlsruhe

KIT Scientific Publishing is a registered trademark of Karlsruhe
Institute of Technology. Reprint using the book cover is not allowed.

www.ksp.kit.edu

Print on Demand 2014

ISSN 1614-5267
ISBN 978-3-7315-0159-6
DOI: 10.5445/KSP/1000037825

Propionic Acid Degradation by Syntrophic Bacteria During Anaerobic Biowaste Digestion

Zur Erlangung des akademischen Grades eines

DOKTOR-INGENIEURS

von der Fakultät für
Bauingenieur-, Geo- und Umweltwissenschaften
des Karlsruher Instituts für Technologie (KIT)

genehmigte

DISSERTATION

von

Dipl.-Ing. Monika Felchner-Żwirełło
aus Gdynia, Polen

<table>
<tr><td>Tag der mündlichen Prüfung:</td><td>08. Februar 2013</td></tr>
<tr><td>Hauptreferent:</td><td>Prof. Dr. rer. nat. habil. Josef Winter</td></tr>
<tr><td>Korreferenten:</td><td>Prof. Dr.-Ing. E.h. Hermann H. Hahn, Ph.D.</td></tr>
<tr><td></td><td>Prof. Dr hab. inż. Jacek Namieśnik</td></tr>
</table>

Acknowledgment

First and foremost, I would like to express my deep sense of gratitude to my supervisor, Prof. Dr. rer. nat. habil. Josef Winter, for inspiring me with his lectures in microbiology during my studies and giving me an opportunity to get to know the syntrophic microorganisms. I appreciate his precious suggestions and guidance during my research work and critical review of the manuscript. I gratefully acknowledge em. Prof. Dr.-Ing. E.h.Hermann H. Hahn, Ph.D. for agreeing to be the Korreferent of my thesis as well as for the valuable discussions and hints he gave me. The heartiest thanks go to Prof. Dr habil. inż. Jacek Namieśnik for believing in me all the time of my doctoral struggle, for his inestimable support and for being my co-referee.

I would like to extend my gratitude to Prof. Dr. Claudia Gallert for the critical verification of my ideas, useful suggestions, and help in technical and non-technical matters.

I owe my sincere appreciation to Prof. Bogdan Zygmunt and Dr inż. Anna Banel for giving me lots of support at the analytical part of this study. I also thank Dr inż. Marek Tobiszewski for his help in administrative affairs at the Gdańsk University of Technology.

To all my colleagues at the Institut für Ingenieurbiologie und Biotechnologie des Abwassers, many thanks for your friendly appearance, full of helpfulness and for unforgettable time I spend in Karlsruhe. Special thanks also go to Dr.-Ing. Mini Bajaj for her help in the laboratory at the very beginning of my research work, to Dr. rer. nat. Daniel Jost for improving my German writing and to Dr. rer. nat. Stephan Bathe for introducing the secrets of FISH to me. I also thank Frau Renate Anschütz for helping me during some laboratory analysis and Frau Rita Seith for the support in administrative matters.

I am also grateful to Politechnika Gdańska (Gdańsk University of Technology) for providing me the financial assistance, and to FAZIT Stiftung for further financial support.

Further thanks go to my friends, who let me believe that I can make the difference and supported me in many ways.

I thank my parents, Jolanta and Bogumił, for "being there" for me at any time and for their love, which made me strong enough to deal with ups and downs during my doctoral studies and in general. *Dziękuję Wam kochani rodzice!* They gave me also the best sister I could imagine. There is no stronger bond in the biosphere than the "sisterhood"; this bond gave me the energy for doing extraordinary things.

Last but not least, I thank my beloved husband Łukasz, who supported me with his advice and his shoulder whenever I needed one (or two). He was and is my inspiration.

Monika Felchner-Żwirełło
Karlsruhe, in summer 2013

About thesis

This dissertation discusses results upon propionic acid degradation under anaerobic conditions in different habitats. Some results published in manuscripts listed below are included in this thesis:

I Felchner-Zwirello M, Winter J, Gallert C (2012) Interspecies distances between propionic acid degraders and methanogens in syntrophic consortia for optimal hydrogen transfer. Appl Microbiol Biotechnol doi: 10.1007/s00253-012-4616-9

II Felchner-Zwirello M, Winter J, Gallert C (2012) Mass spectrometric identification of 13C-labeled metabolites during anaerobic propionic acid oxidation. Chem Biodiv 9:376-384.

Summary

Propionic acid, apart from butyric- and acetic acid, is an important intermediate produced during anaerobic degradation of organic matter and a precursor of a large amount of methane. Its accumulation during the process is however a common problem resulting in a failure of digesters and stagnation phases in biogas production. Such perturbations should be avoided in order to maintain continuous and efficient biogas production. The real reasons for this are not yet fully described and as cultivation of propionate degraders is difficult, the knowledge about syntrophic organisms capable of anaerobic oxidation of propionic acid is restricted. Research allowing better understanding of their behavior and crucial parameters supporting their growth should give a possibility preventing a failure of methanogenesis or drawing a "healing" action plan for reactors at stagnation phases.

During this study, the deeper insight into the process of propionic acid degradation was intended by analyses of the main metabolic pathways in samples containing biomass collected from different mesophilic and thermophilic anaerobic digesters (German full-scale reactors treating biowaste and Indonesian lab-scale reactors treating market waste). The method combining headspace (HS), gas chromatography (GC) and mass spectrometry (MS) was developed and optimized for metabolic pathway analyses with ^{13}C-labelled propionate used as a substrate for microbial cultures. Propionic acid degradation proceeds by the methyl-malonyl-CoA pathway or by the C-6-dismutation pathway, resulting mainly in non- or differently labeled acetic acid production. Mass spectrometry application allowed the detection of ^{13}C-labelling in different metabolites (methyl or carboxyl group of acetic acid) and thus identification of dominant metabolic pathway in biomass samples of different origin. All analyzed microbial consortia degraded propionic acid by the randomized methyl-malonyl-CoA pathway. Furthermore, the developed method may be used to elucidate metabolic pathways when volatile fatty acids (VFAs) are involved.

For the purpose of isolating effectively degrading cultures, several enrichments on propionic acid as a sole carbon source for growth of the cultures of different origin were obtained with and without addition of co-substrates/ electron acceptors such as sulfate, fumarate, pyruvate and crotonate. Fluorescence in-situ hybridization (FISH) was applied for microbial identification of involved microorganisms. Sulfate was determined as the most successful electron acceptor. A tri-culture was isolated from a continuously stirred tank reactor (CSTR) during growth on propionate with sulfate addition. This consisted of three morphologically different organisms: *Syntrophobacter*-like lemon-shaped propionate degraders capable of sulfate reduction; sulfate reducing vibrios and rod-shaped members of the *Eubacteria* group, degraded propionic acid in the presence of sulfate at a high rate of $0.18\,\mathrm{mmol\,l^{-1}\,h^{-1}}$. This culture could be applied as a "healing" agent for digesters in stagnation phases. Additionally a thermophilic co-culture containing hydrogenotrophic methaogens and sulfate reducing *Syntrophobacter*-related propionate degraders was isolated from an industrial scale thermophilic reactor treating biowaste. This co-culture degraded propionate with a rate of $0.06\,\mathrm{mmol\,l^{-1}\,h^{-1}}$.

The ability of propionate degradation by organisms from North Sea sediments and rumen liquid was also investigated. Obtained degradation rates were $0.004\,\mathrm{mmol\,l^{-1}\,h^{-1}}$ and $0.02\,\mathrm{mmol\,l^{-1}\,h^{-1}}$, respectively. The values were compared with those determined for mesophilic and thermophilic cultures, as well as with values available in the literature.

The influence of interspecies distances on hydrogen transfer was investigated in dependence of spatial distances between propionate degraders and methanogens in the enrichment culture obtained from the mesophilic CSTR. The FISH technique combined with epifluorescent microscopy equipped with ApoTom function, as well as 3D image analyses were applied for distance measurements. Additionally, biovolume fraction analysis was performed to describe the population shifts during growth on propionate. The distances between propionate degraders and methanogens were measured with software for microscopic image analysis. They decreased from $5.30\,\mu\mathrm{m}$ to $0.29\,\mu\mathrm{m}$, which caused an increase of the maximum possible hydrogen flux. As the flux was always higher than the hydrogen formation and consumption rate, reducing the interspecies distances by aggregate formation was advantageous in the examined ecosystem.

The research done during this study should give an overall description of syntrophic interactions between the main propionate degraders and the involved methanogens or sulfate reducing bacteria (SRB) and allow pointing out the crucial parameters assuring optimal conditions for their growth. Hence, a mechanism of possible actions avoiding propionate accumulation during anaerobic digestion is suggested (if pH adjustment is not sufficient). This could include interspecies distances minimization for efficient removal of degradation products (hydrogen/ formate and acetate that are toxic for propionate degraders at elevated concentrations) by e.g. aggregation, the use of supporting material for biofilm formation or via the availability of reducing agents. Furthermore, the HS-GC-MS method was developed and successfully applied as a tool for metabolic pathway identification involving the specific carbon distribution within metabolites. Additionally, the isolated tri-culture may be applied in anaerobic digesters as a "healing" agent during failure of the biogas production process by bioaugmentation.

Zusammenfassung

Propionsäure ist neben Butter- und Essigsäure ein wichtiges Zwischenprodukt, welches während des anaeroben Abbaus von organischem Material entsteht. Die Propionsäure kann durch einige Bakterien weiter zu Methan umgesetzt werden. Es kann aber auch zur Propionsäure Anreicherung und Akkumulation im Verlauf anaerober Prozesse kommen, was z.B. in Biogasanlagen eine Biogasproduktionstagnation verursachen würde. Um die optimale und kontinuierliche Biogasproduktion in einer Anlage zu sichern, sollten solche Perturbationen verhindert werden. Allerdings wurden die wirklichen Gründe für dieses Phänomen noch nicht gefunden. Da die Kultivierung von "Propionatabbauern", aufgrund ihres langsamen Wachstums, schwierig ist, ist das Wissen über diese syntrophen Organismen eingeschränkt. Die Forschung zum besseren Verständnis des Verhaltens und die Bestimmung optimaler Wachstumsparameter dieser Organismen sollte eine Möglichkeit geben, einen Ausfall der Methanogenese in Reaktoren zu verhindern oder sogar einen "Problem-Lösungs-Plan" für Reaktoren in Stagnationsphasen zu entwerfen. In dieser Studie wurde der Prozess des Propionsäureabbaus durch Analysen der metabolischen Abbauwege in verschiedenen Biomasse-Proben genauer untersucht. Die Biomasse, die in den jeweiligen Proben gebildet wurde, wurde von vier verschiedenen mesophilen und thermophilen anaeroben Reaktoren gesammelt. Zum einen von zwei deutschen industriellen Biomüllreaktoren und zum anderen von zwei Marktabfall-behandelnden Labor-Reaktoren aus Indonesien. Die Kombination von Headspace (HS), Gas Chromatographie (GC) und Massenspektrometrie (MS) wurde als Methode für die Analyse von metabolischen Abbauwegen der Bakterien, mit Hilfe von ^{13}C-markiertem Propionat als Substrat für das mikrobielle Wachstum, entwickelt und optimiert. Die anaerobe Oxidation von Propionat geschieht entweder durch den Methyl-Malonyl-CoA-Weg oder den C-6-Dismutase-Weg. Das Endprodukt der metabolischen Wege ist hauptsächlich Essigsäure, welche entweder keine oder eine unterschiedliche ^{13}C-Markierung trägt. Die Anwendung der MS ermöglichte die Differenzierung der unterschiedlichen ^{13}C-markierten Metabolite (Markierung bei der Methyl- oder bei der Karboxylgruppe der Essigsäure) und auf diese Weise die Identifizierung des dominierenden metabolischen Abbauweges für die Biomasseproben unterschiedlicher Herkunft. Es konnte herausgefunden werden, dass alle analysierten mikrobiellen Konsortien die Propionsäure durch den randomisierten Methyl-Malonyl-CoA-Weg abbauen. Weiterhin kann die in dieser Studie entwickelte Methode möglicherweise auch für die Bestimmung der beteiligten flüchtigen Fettsäuren (VFAs) angewendet werden.

Einige Anreicherungen der Biomasse unterschiedlichen Ursprungs wurden mit Propionsäure, als einzige Kohlenstoffquelle für mikrobielles Wachstum, entweder mit oder ohne Zugabe von Co-Substraten/Elektronenakzeptoren (wie Sulfat, Fumarat, Pyruvat oder Crotonat) kultiviert. Die Isolierung von effektiven Propionat-abbauenden Kulturen war das Ziel dieses Ansatzes. Die Fluoreszenz *in-situ* Hybridisierung (FISH) diente als Identifikationsmethode von beteiligten Organismen. Sulfat wurde als der am häufigsten verwendete Elektronen-akzeptor bestimmt. Eine "Dreifachkultur" wurde aus einem Rührkessel Reaktor (CSTR) während des Wachstums mit Propionat und zugesetztem Sulfat isoliert. Diese bestand aus drei morphologisch unterschiedlichen Bakterien: Einer oval-förmigen Syntrophobacter-ähnlichen Spezies, welche Propionat und Sulfat abbauen konnte, Sulfat-reduzierenden Vibrien und stäbchenförmigen Mitgliedern der Eubacteriaceae. Eine Propionsäure Abbaurate von $0.18 \, \mathrm{mmol \, l^{-1} \, h^{-1}}$ wurde für diese Anreicherungskultur bestimmt, was sie als möglichen Kandidaten für die Anwendung als "Heilungsmittel" in einer Biogasanlage, die sich in der Stagnationsphase befindet prädestiniert. Zusätzlich wurde eine thermophile Co-Kultur von einem thermophilen Biomüllreaktor isoliert, welche hydrogenotrophe Methanbakterien und Sulfat-reduzierende Syntrophobacter-ähnliche Propionatabbauer einschließt. Diese thermo-phile Kultur hat Propionsäure mit einer Rate von $0.06 \, \mathrm{mmol \, l^{-1} \, h^{-1}}$ abgebaut.

Die Fähigkeit des Propionsäureabbaus von Organismen aus Nordseesedimenten und aus der Pansenflüssigkeit von Wiederkäuern wurde ebenfalls erforscht. Die Abbauraten von $0.004 \, \mathrm{mmol \, l^{-1} \, h^{-1}}$ und $0.02 \, \mathrm{mmol \, l^{-1} \, h^{-1}}$ wurden jeweils bestimmt. Ein Vergleich dieser Werte mit Literaturwerten wurde durchgeführt und diskutiert.

Als weiterer Teil dieser Arbeit wurden die räumlichen Abstände zwischen Propionatabbauern und Methanbakterien in einer mesophilen Anreicherungskultur aus dem CSTR auf den Wasserstoff-Transfer zwischen den Spezies untersucht. Die FISH-Technik wurde kombiniert mit der Verwendung eines Epifluoreszenzmikroskopes mit ApoTome-Funktion damit eine 3D-Bildanalyse für Distanzmessungen zwischen den Bakterienzellen möglich war. Zusätzlich wurde eine Analyse des Biovolumenanteils verwendet um die Bestandsveränderungen der Bakterien während des Wachstums mit Propionat zu beobachten und zu analysieren. Die mit Hilfe der Imagesoftware gemessenen Distanzen zwischen den Zellen nahmen von $5.30\,\mu m$ auf $0.29\,\mu m$ ab. Diese Abnahme begünstigte die Zunahme des maximal möglichen Wasserstoffflusses. Da der berechnete H_2-Fluss jeweils höher als die Wasserstoffbildungsrate und auch die Wasserstoffverbrauchsrate war, kann eine Verringerung der interzellulären Abstände durch Aggregatformation von syntrophen Bakterienzellen vorteilhaft für das untersuchte Ökosystem sein.

Die Forschungsarbeit, die während dieser Studie erfolgt ist, beschreibt die syntrophen Interaktionen zwischen den Propionat-abbauenden Bakterien und den beteiligten Methan- oder Sulfat-reduzierenden Bakterien (SRB). Außerdem wurden die entscheidenden Parameter, die für optimale Wachstumsbedingungen notwendig sind, untersucht und verifiziert. Eine Möglichkeit einer Propionatanreicherung bzw. Akkumulation (wenn eine manuelle pH-Korrektur nicht ausreichend ist) entgegen zu wirken wurde somit vorgeschlagen. Diese Möglichkeit könnte eine Interspezies-Abstandsminimierung für eine leistungsfähige Entnahme von Abbauprodukten (wie Wasserstoff/ Formiat und Acetat, die für Propionatabbauer bei erhöhten Konzentrationen giftig sein können) einschließen. Durch die Distanzminimierung zwischen den Zellen (z.B. durch Aggregation), oder auch die Anwendung von Trägermaterialien für eine stärkere Biofilmbildung oder die erhöhte Verfügbarkeit von Reduktionsmitteln lässt sich die Methanproduktion steigern. Außerdem, wurde in dieser Arbeit die HS-GC-MS Methode, welche die Ermittlung der spezifischen Kohlenstoffdistribution bei den Metaboliten beinhaltet, entwickelt und als Instrument zur Differenzierung von metabolischen Abbauwegen der Propionatabbauer erfolgreich angewendet. Weiterhin, kann die isolierte Dreifachkultur als sogenanntes "Heilungsmittel" durch Bioaugmentation, für Anlagen die sich in einer Biogasproduktionsausfallphase befinden, verwendet werden.

Table of contents

Abbreviations

μ	: Specific growth rate
AM	: Axial modulation
APHA	: American Public Health Association
ASTM	: American Society for Testing and Materials
ATP	: Adenosine triphosphate
BESA	: 2-bromoethane-1-sulfonic acid
CARD	: Catalyzed reporter deposition
CAS	: Chemical Abstracts Service
CoA	: Coenzyme A
CSTR	: Continously stirred tank reactor
CV	: Coefficient of variation
Cy3	: Cyanine 3
Cy5	: Cyanine 5
DAPI	: 4',6-diamidino-2-phenylindole
DGGE	: Denaturing gradient gel electrophoresis
DNA	: Deoxyribonucleic acid
DSM	: Deutsche Sammlung von Mikroorganismen (German Collection of Microorganisms)
DSMZ	: Deutsche Sammlung von Mikroorganismen und Zellkulturen (Germ an Collection of Microorganisms and Cell Cultures)
DY415	: Dyomics GmbH fluorescent dye
EEG	: Erneu bare-Energien-Gesetz (Renewable energy sources act)
EI	: Electron ionization
EPA	: US Environmental Protection Agency
$FADH_2$	: Flavin adenine dinucleotide
FAM	: 6-carboxylfluorescein
FFAP	: Free fatty acids phase
FID	: Flame ionization detector
FISH	: Fluorescence *in-situ* hybridization
FITC	: Fluorescein isothiocyanate
GC	: Gas chromatography

HPLC   : High performance liquid chromatography
HS   : Headspace
IBA   : Institute of Biology for Engineers and Wastewater Treatment
IC   : Ion chromatography
IHT   : Interspecies hydrogen transfer
ISO   : International Standard Organization
ITB   : Institute of Technology Bandung
LIPI   : Indonesian Institute of Sciences
LOD   : Limit of detection
LOQ   : Limit of quantification
m/ z   : Mass/ charge ratio
M   : $mol\,l^{-1}$
MAR   : Microaudiography
MeOH   : Methanol
mM   : $mmol\,l^{-1}$
MS   : Mass spectrometry
$NADH_2$: Nicotinamide adenine dinucleotide
NIST   : National Institute of Standards and Technology
p.a.   : pro analisi (pro analysis)
PBS   : Phosphate buffered saline
PC   : Personal computer
PFA   : Paraformaldehyde
ppm   : Part per million
RNA   : Ribonucleic acid
rpm   : Rotations per minute
rRNA   : Ribosomal RNA
SCAN   : Full spectrum scanning
SDS   : Sodium dodecyl sulfate
SIM   : Selected ion monitoring
SPME   : Solid phase microextraction
SRB   : Sulfate reducing bacteria
SSU   : Small-subunit
TCD   : Thermal conductivity detector
t_d : Doubling time
Tris   : Tris(hydroxymethyl)aminomethane
TRITC   : Tetramethylrhodamine 5-isothiocyanate
UASB   : Up-flow anaerobic sludge blanket reactor
v/ v   : volume per volume

vi

VFAs	: Volatile fatty acids
VSS	: Volatile suspended solids
w/ v	: Weight per volume
w/ w	: Weight per weight
$\Delta G^{\circ\prime}$	: Gibbs free energy of reaction at standard conditions and pH = 7

Further abbreviations are explained in text.

List of figures

List of tables

1 Introduction

Anaerobic biowaste degradation is known as a possibility for utilization of organic waste fractions, fertilizer production and a source of biogas. Biogas consists of about 65 % methane, which is an energy source (McInerney et. al, 2008) and means also a renewable energy source (according to Erneuerbare-Energien-Gesetz, EEG). The approach is gaining importance to meet the goals of European and national policies and instructions described in several acts concerning energy production, agriculture and the environment (Stronach et al. 1986, Inanc et al. 1999). Increasing energy demand of mankind and exploiting fossil fuels supply forced the search for alternative energy sources. The anaerobic digestion of biological wastes seems to be one solution, as is not only considered to be environmentally friendly, but also an economically beneficial way of methane production (Chandra et al. 2012). According to Deublein and Steinhauser (2008), the energy output/ input ratio of this biomass conversion into energy is the most efficient one in comparison with other biological and thermo-chemical technologies. Application of anaerobic digestion for organic waste treatment has been increasing annually by 25 % in recent years (Appels et al. 2011). The German Renewable Energy Sources Act 2012 (EEG) set the renewable energy sources supply share in the total gross final energy consumption by the end of the year 2020 to 18 % or to 35 % share in electricity supply (German Federal Ministry for the Environment, Nature Conservation and Nuclear Safety, 2012). Lebun et al. (2011) forecasted the number of biogas producing plants in Germany at the end of the year 2011 to 7000 with an electrical capacity of 2.6 GW. These facts show that biomethanation is a process with an extensive potential, however, its complexity concerning participating microbial consortia makes it hard to conduct without stagnation phases. The microflora is not yet fully researched and the cultivation of these organisms, if at all possible, is difficult (Klocke, 2011). Knowing the main participants of each step is fundamental to find the optimal parameters for achieving the highest efficiency of biogas production. A very common disturbance phase found for biogas production is the accumulation of volatile fatty acids (VFAs) especially of propionic acid and a rapid pH decrease (Schievano et al. 2010). On the other hand fatty acids, including propionic acid are considered to be precursors of large methane amount (Ariesyday et al. 2007) and hence, are important intermediates in biogas production. The studies on accumulation reasons especially for propionate are demanded as the reasons for disturbances in the process are not yet fully examined. Fluctuations of propionate concentrations during anaerobic processes are the reason for low performance of methane production. Furthermore, large amounts of propionate may inhibit the process. To solve the problem, its origin has to be known. Research on the main propionate degraders should allow drawing a procedure for the optimal biogas plant operation without process failure.

1.1 Methods of organic waste treatment

Organic wastes can be classified according to their origin: agricultural, municipal (including market waste), domestic or industrial. Although the amounts produced vary from country to country, generally the largest magnitude is delivered by the first two mentioned sectors. There are three possibilities of solid wastes treatment: physical (e.g. volume and size reduction), chemical (gasification, combustion, pyrolysis) and biological (anaerobic digestion, aerobic composting) (Tchobanoglous et al. 1993). Generally it is said, that the most appropriate methods for treatment of organic wastes are the biological methods. Composting is the art of aerobic biological conversion of organic waste resulting in production of compost (Tchobanoglous et al. 1993).

Anaerobic digestion results in biogas production, which consists mainly of methane, carbon dioxide and hydrogen sulfide if the process is conducted in the absence of oxygen and at neutral pH (optimal pH: 6.7-7.4) (Bryant 1979). As the conversion occurs thanks to microorganisms' contribution, a "new biomass" is also produced, which can be used as a fertilizer. Spreading of excessive sludge e.g. on agricultural lands is a common practice (Gong et al. 1997). Considered anaerobic degradation process involves four stages: hydrolysis, acidogenesis, acetogenesis and methanogenesis (Fig. 1.1). Each step is performed by different microbial species. Sustaining a well-balanced microbial ecosystem is thus crucial for the process and approaches like co-digestion of different organic wastes (with addition of wastewater sludge (Hahn and Hoffmann 1992) are commonly applied to achieve that. Products obtained in one phase of conversion are used as substrates for the next one as shown in Figure 1. Most of the reactors operate under mesophilic (20 - 40 °C) or thermophilic conditions (45 - 60 °C) (Borja 2011). Psychrophilic (below 20 °C degradation of organic wastes is relative rarely applied, but in low temperature areas like the Himalayas or the Alpes in Switzerland, where the operation of conventional digesters is too expensive, psychrophilic digesters were installed (Dhaked et al. 2010).

Hydrolysis is the step at which biopolymers such as lipids, proteins and carbohydrates are degraded into higher fatty acids, glycerol, monomeric carbohydrates, amino acids and alcohols. The conversion is possible thanks to extracellular enzymes such as lipases, proteases, amylases and cellulases (Fox and Pohland 1994). Hydrolytic bacteria like *Clostridium sp.*, *Proteus vulgaris*, *Peptoccocus sp.*, *Bacteriodes sp.*, *Bacillus sp.*, *Vibrio sp.*, *Acetovibrio cellulitics*, *Staphyllococcus sp.* and *Microccocus sp.* are involved (Stronach et al. 1986).

Acidogenesis (the production of fatty acids from monomers after hydrolytic cleavage) is also known as fermentation. Products obtained during hydrolysis are here converted mainly into primary volatile fatty acids, including propionic acid. Acidogenic bacteria (like for instance *Lactobacillus sp.*, *Staphylococcus sp.*, *Escherichia sp.*, *Desulfobacter*, *Desulfomonas*, *Veillonella sp.*, *Selenomonas sp.*, *Streptococcus sp.*, *Clostridium sp.* and *Eubacterium limosum*) are the participants of this digestion step (Tchobanoglous et al. 1993).

Acetogenesis involves acetogenic bacteria, which produce acetic acid from previously obtained longer-chain VFAs, and homoacetogenic bacteria, which use hydrogen and carbon dioxide for acetate formation (Gerardi 2003). *Clostridium sp.*, *Syntrophomonas wolfei* and *Syntrophobacter wolinii* are examples of acetogens (Stronach et al. 1986). Acetogenesis of volatile fatty acids requires a low hydrogen partial pressure, as discussed later on.

Methanogenesis is the final step of anaerobic digestion where microorganisms from the *Archaea* domain convert either acetic acid into methane (aceticlastic methane bacteria like e.g. *Methanosarcina sp.*, *Methanosaeta sp.*) or carbon dioxide and hydrogen (hydrogenotrophic methane bacteria like e.g. *Methanobacterium sp.*, *Methanobrevibacterium sp.*, *Methanospirillum sp.*). Methane production can also occur by anaerobic digestion of methanol, formate or methylamines (Boone et al. 1993) Hydrogenotrophic methanogens are important for maintenance of a low hydrogen partial pressure which is a prerequisite for acetogenesis.

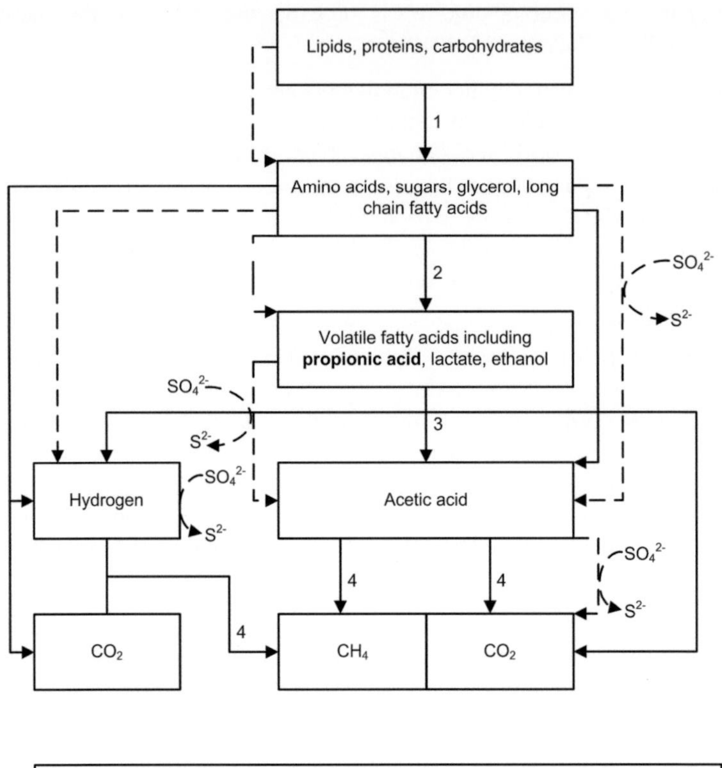

Fig. 1.1: Schematic representation of steps of organic matter anaerobic degradation (solid lines) with additionally marked pathways in the presence of sulfate (broken lines) according to Muyzer and Stams (2008).

1.2 Volatile fatty acids in anaerobic digestion

As described in the previous section, VFAs are produced during the second step of anaerobic biowaste digestion- acidogenesis. According to different literature sources, compounds of this group are low molecular weight aliphatic carboxylic acids with a chain length of 2 to 5 (Chen et al. 1994, Cruvys et al. 2002) or 6 (Abalos et al. 2000, Larreta et al. 2006) carbon atoms. They are not only considered to be products of bacterial activity, but also as important carbon sources and hence, are crucial in monitoring the performance of anaerobic digestion. If the rate of their production exceeds the degradation rate, VFAs accumulation is observed- especially of propionic-, butyric- and acetic acid. Such state can result in failure of a process or even in its inhibition (Asinari Di San Marzano et al. 1981, Dröge et al. 2008). Several causes of inhibition of anaerobic digestion has been described in the literature, like elevated level of ammonia, hydraulic or organic overload which often lead to VFAs accumulation, pH drop and digester failure (Schievano et al. 2010). This results in methane production stagnation and must be avoided, but the science does not give clear explanation of reasons for the VFAs accumulation.

Propionic acid is one of the major intermediates in anaerobic digestion and a precursor of about 35 mol% methane (together with acetic acid for approximately 85 % (Lawrence and McCarty 1969). Degradation of propionic acid is also considered to be the rate-limiting factor for the whole process (Vavilin et al. 2003, Ariesyady et al. 2007). Furthermore, the importance of this acid as an intermediate during anaerobic methane production was mentioned in many publications (e.g. Schink 1988, Plugge et al. 1993, Inanc et al. 1999, Liu et al. 1999, Ariesyady et al. 2007, McInerney et al. 2008). However, due to its thermodynamically unfavorable anaerobic oxidation ($\Delta G^{o\prime} = +76.1\,\text{kJ}$), which can not occur spontaneously, it is known as a problematic compound. Similarly unfavorable thermodynamic conditions prevail for n-butyrate ($\Delta G^{o\prime} = +48.1\,\text{kJ}$). Both VFAs require a very low hydrogen partial pressure to shift the reaction stoichiometry towards degradation and hence syntrophic partner organisms (Dong and Stams 1995). Nevertheless, propionate is thought to accumulate easily than acetic- and butyric acid because of its low conversion rate resulting in indirect subjection to methanogenesis (Shin at al. 2010). Moreover, there have been studies which showed that propionic acid has a higher inhibitory effect on methanogenic bacteria than acetic- or butyric acid accumulation (Wang et al. 2009). McCarty and Brosseau (1963) (in Inanc et al. 1999) reported that the tolerable level of butyric acid for methanogenic bacteria was ten times higher than that for propionate. On the other hand some researchers have not noticed any meaningful methane production decrease at propionate concentration of $2750\,\text{mg}\,\text{l}^{-1}$(Pullammanappallil et al. 2001). It shows that the issue itself is recognized, but unfortunately the real reasons for this problem are not yet fully defined and the microbial species description is until now scarce. Despite of this, several researchers proposed some solutions, mainly for the problem of propionate accumulation: adaptation to inhibitory substances (Chen et al. 2008), macro- and micro-nutrients supplementation (Ma at al. 2009), enhanced propionic acid degradation (EPAD) system (Ma at al. 2009 a), predicting optimal parameters by mathematical models (Sbarciog et al. 2010) or introducing acetate enrichment culture (ADEC) into digester

(Lins at al. 2010). The mentioned approaches are however in most cases time consuming and not generally applicable.

Syntrophic associations between involved bacteria are known to be crucial for optimal VFAs degradation and good quality biogas production; however, there is still too little information about them. Still many researchers describe anaerobic degradation of organic matter as a "black box" (Dabert et al. 2002). The need for analyzing microorganisms' communities and their contribution to anaerobic degradation of organic compounds in anaerobic digesters is essential to understand the process and to maintain a stable ecosystem by maintaining suitable conditions (McInerney et al. 2008, Shin et al. 2010).

According to studies showing that VFAs have an inhibitory effect on overall methanogenesis, an optimal ratio between them has to be maintained. This was confirmed by Marchaim and Krause (1993), who proved with several lab-scale thermophilic reactors that were running with different feeding regimes by increasing propionate to acetate ratios, that there was an immediate biogas composition change, namely a CH_4 concentration decrease. That denoted this ratio as an immediate indicator for organic overload. Additionally, Hajaranis and Ranade (1994) studied the influence of VFAs on methanogenic bacteria activity. Methane production inhibition by butyrate, valerate and caproate was tested at acids concentrations in the range from 5 to 20 g l^{-1}. A 90 % inhibitory effect was estimated for all compounds at a level of 20 g l^{-1} in comparison with cultures growing without acids' addition. Furthermore, the i- isomers of analyzed compounds were less harmful than their n-forms.

1.3 VFAs' analytical measurement methods

Scientific literature describes that first measurements of VFAs were performed by titration preceded by analytes isolation through distillation. Samples prepared by means of extraction, thermal desorption and/ or derivatisation are commonly examined by chromatographic methods. These include, gas chromatography (GC), ion chromatography (IC) and high performance liquid chromatography (HPLC). The last two are considered to be demanding, because of the sample purification necessity, the requirement of the addition of certain agents to eliminate other ions interference. As a gas chromatograph equipped with flame ionization detector (FID) can be used for the analysis of non-derivatised VFAs, it is most commonly used. However, due to the need of more precise and less time consuming approaches, solvent-free HS-GC and HS- solid phase microextraction (SPME), as well as SPME (immersing the fiber directly in the sample) modification were developed (Abalos et al. 2000, Cruvys et al. 2002, Boe et al. 2007). Apart from popular FID, detectors such as MS and combination of MS/ olfactometry (O) were used in order to indicate and quantify the VFAs (Zhang et al. 2010). The decision on method selection for VFAs analysis is mainly based on the goal of planned research, demanded results precision, sample matrix, available equipment, the level of analytes in the sample, as well as on its time consumption and connected expenses (Namieśnik and Jamrógiewicz 1998). The use of the GC-FID technique is the most widespread among researchers (e.g. Ma et al. 2009, Schievano et al. 2010, Lozecznik et al. 2012) due to its quite uncomplicated sample preparation requirements, where usually filtration or centrifugation is sufficient. However, choosing an appropriate chromatographic column, which is water-resistant and allows obtaining chromatograms of a good resolution, is not straightforward (Peldszus 2007). FID is commonly applied, because of its detection ability of almost all organic compounds (excluding formate and carbon disulfide) (Sczepaniak 2005) and good sensitivity (0.1 - 10 ng) (Namieśnik and Jamrógiewicz 1998). GC-MS with electron ionization (EI) is a method appropriate for the analysis of compounds of high and medium volatility (Dincer et al. 2006, Sassaki et al. 2008, Zhou et al. 2009). MS as a detector has a sensitivity of 1 - 10 ng at full spectrum scanning (SCAN) mode, and 1 - 10 pg at selected ion monitoring (SIM) mode (Namieśnik and Jamrógiewicz 1998) and additionally the resulting mass spectrum of an analyte includes specific fragmentation ions which characterize detected compound and hence increases the detection reliability.

1.4 The problem of anaerobic oxidation of propionic acid

Syntrophy is "a special case of symbiotic cooperation between two metabolically different types of bacteria which depend on each other for degradation of a certain substrate, typically for energetic reason" (Schink 1997). In other words, it is a dependency between two organisms where one produces metabolites from a certain compound by anaerobic oxidation, which are efficiently removed by the partner. This allows them gaining energy for growth from reactions that are thermodynamically unfavorable or even enderogonic under standard conditions. Anaerobic oxidation of propionic acid is an example of such a reaction, and that is probably the main reason for the perturbations during anaerobic digestion of organic wastes. This reaction results in acetate and hydrogen or formate production. Although the interspecies transfer of

Table 1.1: Tolerable concentrations of propionic acid during biowaste degradation

Tolerable propionate $[g\,l^{-1}]$	Reference
0.8	Mosche and Jordening 1998
1.0	McCarthy and Brosseau 1963 in Inanc et al. 1999
1.0 - 2.0	Barredo and Evison 1991
2.5	Satoto 2009
> 3	Pullammanappallil et al. 2001

formate has a significant meaning (Thiele and Zeikus 1988, Boone et al. 1989), H_2 transfer is usually associated with syntrophic relations between organisms (Schink 1997). Because both, hydrogen or formate are present at low concentrations, it is hard to differentiate which one plays the main role (de Bok et al. 2004).

It has been reported, that propionic acid is one of the compounds (apart from acetic-, butyric acid and alcohols), which accumulate during anaerobic digester overload or inhibition (Gallert and Winter 2008). Unstable conditions in methane reactors often lead to an increasing hydrogen partial pressure accompanied by propionate and sometimes butyrate accumulation and a decreasing performance of methane production (Inanc et al. 1999, Gallert and Winter 2008), which finally will lead to failure of methanogenesis, if no counteractions are undertaken.

Concentration level of propionic acid causing biowaste degradation inhibition may vary. It depends on the anaerobic reactor type and the type of treated waste, but generally it is assumed that $1.5\,g\,l^{-1}$ is a maximal concentration at which a stable process still might be maintained. The span of propionate tolerable levels determined by several researchers is scattering within the range of $0.8 - 3.0\,g\,l^{-1}$ (Table 1.1).

The role of hydrogen and formate

Oxidation of propionate is associated with the production of hydrogen and can be exerogonic only when the partial pressure of hydrogen is kept below 10^{-4}bar. This indicates, that propionate degradation is only proceeding with hydrogen consuming participant- chemical substance or syntrophic partner (Vavilin et al. 2003, de Bok et al. 2004, Shigematsu et al. 2005). Methane bacteria, which convert H_2 into CH_4, are usually pointed out as such partners (Vavilin et al. 2003, de Bok et al. 2004) The free Gibbs energy of methane production from H_2 and CO_2 is negative only when the partial pressure of hydrogen is above the value of 10^{-6} bar. It means that propionic acid conversion is only possible in the hydrogen partial pressure range between 10^{-6} bar and 10^{-4} bar or 0.1 - 10 Pa (Schink 1988, Ariesyady et al. 2007, Larreta et al. 2006). These values determine the area of the so-called thermodynamic window (Fig. 1.2). The reaction stochiometries of propionic acid degradation and methane formation and their $\Delta G^{\circ\prime}$ values are shown below ((1.1) - (1.3)).

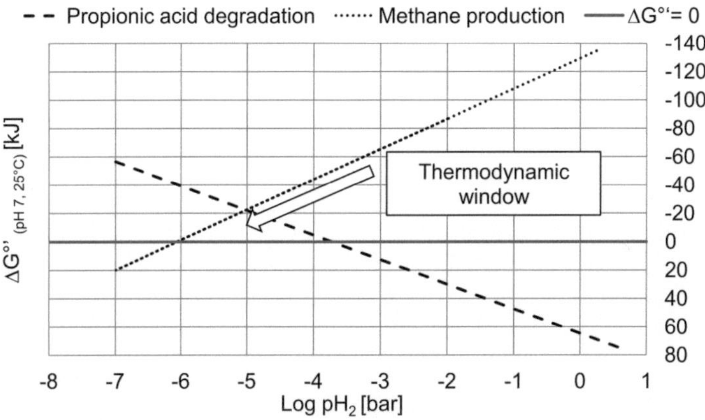

Fig. 1.2: Gibbs free energy in relation to hydrogen partial pressure for methane production from propionic acid oxidation.

$$CH_3CH_2COO^- + 3\,H_2O \longrightarrow CH_3COO^- + HCO_3^- + H^+ + 3\,H_2 \qquad (1.1)$$
$$\Delta G^{\circ\prime} = +76.1\,\mathrm{kJ/reaction}$$

$$CH_3COO^- + H_2O \longrightarrow CH_4 + HCO_3^- \qquad (1.2)$$
$$\Delta G^{\circ\prime} = -31\,\mathrm{kJ/reaction}$$

$$4\,H_2 + HCO_3^- + H^+ \longrightarrow CH_4 + 3\,H_2O \qquad (1.3)$$
$$\Delta G^{\circ\prime} = -135.6\,\mathrm{kJ/reaction}$$

Because of a positive $\Delta G^{\circ\prime}$, propionic acid degradation depends on successful hydrogen transfer between participating organisms. It has been stated that hydrogen interspecies transfer, together with formate (as some species of hydrogenotrophs also use formate) play a crucial role for the optimal degradation of propionate. Studies with *S. fumaroxidans* showed the involvement of at least three hydrogenases and two formate dehydrogenases in its metabolism (de Bok et al. 2002), indicating that both, hydrogen and formate may serve as electron carriers during propionate oxidation. However, their relative contribution to syntrophic degradation of propionate is difficult to determine and, hence, not sufficiently clear (Stams and Plugge 2009). There are two known metabolic pathways of anaerobic propionate oxidation: The methyl-malonyl-CoA pathway, where from one mole of propionate, one mole of acetate and one mole of carbon dioxide are formed (Plugge et al. 1993), and the reducing equivalents at the level of $NADH + H^+$, $FADH_2$ and ferredoxin appear from conversion of succinate into fumarate, malate to oxaloacetate and pyruvate to acetyl-CoA (Stams 1994). The second pathway is the C-6-dismutation pathway (de Bok et al. 2001), where from two moles of propionate, three moles of acetate are produced with n-butyrate formation as an intermediate metabolite in the conversion. A schematic representation of both pathways is shown in the Materials and methods section (Figures 3.7 and 3.8).

Hydrogen (1.3) and/ or formate utilization (1.4) for methane production or sulfate reduction (1.5) is the driving force of anaerobic propionate oxidation. In anaerobic digesters methane formation keeps the hydrogen partial pressure pH_2 low, what allows propionate oxidation (Thauer et al. 1977).

$$4\,HCOO^- + H_2O + 4\,H^+ \longrightarrow CH_4 + 3\,HCO_3^- + 3\,H^+ \qquad (1.4)$$
$$\Delta G^{\circ\prime} = -130.4\,kJ/reaction$$

$$4\,H_2^- + SO_4^{2-} + 2\,H^+ \longrightarrow H_2S + 4\,H_2O \qquad (1.5)$$
$$\Delta G^{\circ\prime} = -154\,kJ/reaction$$

The bioenergetic conditions of the anaerobic metabolism are controlled by the specific activity of hydrogen-oxidizing methanogens (Dwyer et al. 1988). This phenomenon was named by Bryant et al. (1967) and Wolin (1975) as interspecies hydrogen transfer (IHT), but formate apparently can also be used in interspecies electron flow (Thiele and Zeikus 1988, Bleicher and Winter 1994, Stams and Dong 1995). If the diffusion constant and solubility of hydrogen and formate were compared, it can be expected from Fick´s diffusion law, that formate transfer is the favored mechanism when interspecies distances are high, while hydrogen transfer becomes more favorable when interspecies distances are low (de Bok et al. 2004). Recent biochemical studies and genome analyses of pure cultures of syntrophic volatile fatty acid degraders indicated, that hydrogen and formate transfer occurred simultaneously (Müller et al.

2010). (1.6) and (1.7) represent possible reactions of formate production from propionic acid.

$$CH_3CH_2COO^- + 2\,H_2O \longrightarrow CH_3COO^- + HCOO^- + H^+ + 2\,H_2 \qquad (1.6)$$
$$\Delta G^{\circ\prime} = +75\,kJ/reaction$$

$$CH_3CH_2COO^- + 2\,HCO_3^- \longrightarrow CH_3COO^- + 3\,HCOO^- + H^+ \qquad (1.7)$$
$$\Delta G^{\circ\prime} = +72.4\,kJ/reaction$$

Possible hydrogen sinks

Propionic acid degraders are syntrophs which live in close neighborhood with a hydrogen/ formate consuming partner, as described above, or require hydrogen incorporating chemical substance, a co-substrate. Most popular and probable occuring in the natural environment of anaerobic oxidation are the following chemicals that can serve as electron acceptors: sulfate, fumarate, pyruvate and crotonate.

Sulfur is present on the Earth in rocks, sediments, as component of pyrite or gypsum and in seawaters as soluble sulfate. Microorganisms are crucial for the global sulfur cycle, as they take this element up in the form of sulfate and release it as sulfide during organic matter decomposition. Sulfate occurs in anoxic or anaerobic habitats like e.g. rice field soil (Stubner 2004), mud volcanoes (Stadnitskaia et al. 2005), plants rhizosphere (Hines et al. 1999), marine sediments (Bowles et al. 2011) or anaerobic digesters treating sulfate rich wastewater (from e.g. molasses or production industry (Percheron et al. 1999)). The presence of sulfate reducing bacteria using SO_4^{2-} in the form of terminal electron acceptor for growth in mentioned ecosystems was confirmed by several studies (Muyzer and Stams 2008). Most propionic acid oxidizing organisms are also able to reduce sulfate (Table 1.2), and thus are a special case of sulfate reducing bacteria. Propionic acid degradation (1.1) coupled with SO_4^{2-} conversion to sulfide (1.5) allows these organisms to grow, as the $\Delta G^{\circ\prime}$ for this coupled reaction diminishes and makes the process exergonic. Such combination is energetically more beneficial than syntrophic oxidation of propionate and enhances the growth of considered bacteria. The activity of sulfate reducing bacteria can however be the reason for imbalances of methanogenesis in anaerobic environment e.g. during biowaste digestion, as they are suspected to compete for hydrogen and, in some cases, for acetate with methanogens (Muyzer and Stams 2008).

The other three mentioned possible hydrogen sinks (fumarate, pyruvate, crotonate) are metabolites of degradation pathways active during biowaste digestion, including propionate oxidation. Due to their appearance and chemical structure it is probable that syntrophs could use them for growth as a terminal electron acceptor. Fumarate has been proved to serve as such for *Syntrophobacter fumaroxidans* (Harmsen et al. 1998). Fumarate in this case was a successful agent replacing methanogens (as syntrophic partner) during bacterial growth on propionate in the experiment with "advanced mesophilic enrichment cultures" (propionate oxidation was observed with inhibited methanogenesis and fumarate addition) (Stams et al. 1993). As the step of succinate oxidation to fumarate in the metabolic pathway of propionate oxidation is difficult $\Delta G^{\circ\prime} = +86.2\,\mathrm{kJ/mol}$, it might be more convenient for considered microorganisms to omit this step if fumarate is available (Stams et al. 1993).

Syntrophobacter wolinii can grow fermentatively on pyruvate (Stams et al. 1993, Wallrabenstein et al. 1994), as well as *S. fumaroxidans* (Hermie et al 1998) and *S. pfenigii* (Wallrabenstein et al. 1995). Some syntrophs were also reported to be able to grow with crotonate (Beaty and McInerney 1987, Wallrabenstein et al. 1995, Zhao et al. 1990), generally butyrate or benzoate degraders. These substrates did not serve as external electron acceptor for syntophic growth, but their chemical structure might allow them serving such role.

1.4.1 Bacterial species capable of propionic acid oxidation

The diversity of microorganism groups taking part in each step of biowaste conversion to methane and carbon dioxide make the whole process hard to manage and difficult to describe. Syntrophic associations between involved bacteria are known to be crucial for good quality biogas production; however, there is still too little information about them. The need for analyzing microorganisms' communities in anaerobic digesters is essential to understand the process and to facilitate a stable ecosystem by maintaining optimal conditions (McInerney et al. 2008, Rittmann et al. 2008, Talbot et al. 2008, Shin et al. 2010). If propionic acid degradation is considered, there is quite a small number of bacterial species described that are capable of this. Due to unfavorable thermodynamic of propionate conversion, most of them oxidize it in a co-culture with a hydrogen/ formate utilizing partner or with a suitable chemical compound as an electron acceptor. Syntrophic propionate degraders are members of deltaproteobacteria of the *Syntrophobacteriales* order, and of the *Clostridiales* order of *Peptococcaceae*.

Known species from deltaproteobacteria such as *Syntrophobacter pfennigii, Syntrophobacter fumaroxidans, Syntrophobacter sulfatireducens, Syntrophobacter wolinii* and *Smithella propionica* have been previously described in co-cultures with *Methanospirillum hungatei, Desulfovibrio* or in axenic cultures with co-substrates such as sulfate, sulfite, fumarate or growing axenically on pyruvate or fumarate (Boone and Bryant 1980, Harmsen et al. 1993, Wallrabenstein et al. 1995, Harmsen et al. 1998, Liu et al. 1999, de Bok et al. 2001, Chen et al. 2005). *Syntrophobacter* genus members are physiologically and phylogenetically related to sulfate reducing bacteria (Stams and Plugge 2009) and can couple propionate oxidation with sulfate reduction. *Smithella propionica* is the only exception among *Syntrophobacterales* which cannot reduce sulfate (McInerney et al. 2008). There are also members of *Firmicutes* described in a co-culture with methanogenic partners (*M. hungatei, Methanobacterium thermoautotrophicum*) like *Pelotomaculum schinkii, Pelotomaculum thermopropionicum, Desulfotomaculum thermocisternum, Desulfotomaculum thermobenzoicum* subsp. *thermosyntrophicum* (Nilsen et al. 1996, Plugge et al. 2002, Imachi et al. 2002, de Bok et al. 2005). The mesophilic propionate degraders are of the genera *Syntrophobacter, Smithella* and *Pelotomaculum*. Thermophilic ones are *Desulfotomaculum* and *Pelotomaculum* (Vavilin et al. 2003, Ariesyday et al. 2007). The organisms together with their syntrophic partners and possible co-substrate or substrates are described in Table 1.2.

Syntrophobacter wolinii was the first described organism among syntrophic propionic acid degraders obtained in a co-culture with *Desulfovibrio sp.* from an anaerobic digester treating municipal sewage (Boone and Bryant 1980).

Specific growth rates for considered organisms are very low in comparison to e.g. *Eschericha coli*, whose doubling time equals 20 minutes, and are in the range 0.02 d^{-1} - 1.66 d^{-1}, which will be discussed in section 5.4.1, Table 5.4. This means time consuming cultivation of mentioned species makes the research with them difficult.

Smithella propionica is Gram- negative, spore forming organism (Liu et al. 1999) degrading propionic acid with intermediate butyrate production via the C-6-dismutation pathway (de Bok et al 2001). All other representatives of propionate oxidizers use the methyl-malonyl-CoA pathway (Plugge et al. 1993).

Table 1.2: Chosen propionic acid degraders and their partners including the possibility of sulfate reduction coupled to propionate degradation and possible substrate for axenic growth (co-substrate means external electron acceptor replacing methanogenic partner other than sulfate).

Propionic acid degraders	Syntrophic partner / co-substrate/ substrate	Sulfate reduction	Reference
Syntrophobacter fumaroxidans	*Methanospirillum hungatei*/ fumarate/ fumarate, pyruvate	+	Harmsen et al. 1998
Syntrophobacter pfennigii	*Methanospirillum hungatei*/ sulfite, thiosulfate/ lactate	+	Wallrabenstein et al. 1995
Syntrophobacter sulfatireducens	*Methanospirillum hungatei*	+	Chen et al. 2005
Syntrophobacter wolinii	*Methanospirillum hungatei, Desulfovibrio sp.*/ pyruvate	+	Boone and Bryant 1980, Wallrabenstein et al. 1994,
Smithella propionica	*Methanospirillum hungatei*/ crotonate, butyrate	-	de Bok et al. 2001, Liu et al. 1999, Wallrabenstein et al. 1994
Pelotomaculum schinkii	*Methanospirillum hungatei*	-	de Bok et al. 2005
Pelotomaculum thermopropionicum	*Methanothermobacter thermautotrophicum* / - / fumarate, pyruvate	-	Imachi et al. 2002
Desulfotomaculum thermocisternum	*Methanobacterium thermoautotrophicum* / - / fumarate, pyruvate	+	Nilsen et al. 1996
Desulfotomaculum thermobenzoicum subsp. *thermosyntrophicum*	*Methanobacterium thermoautotrophicum* / - / fumarate, pyruvate	+	Plugge et al. 2002

1.4.2 Organization of syntrophic bacteria

Bacteria that degrade propionate syntrophicaly have to share the energy for growth with their hydrogen/formate consuming partner. Complete mineralization of this acid requires three participants: an acetogenic bacterium and two methanogens, one being hydrogenotrophic and one aceticlastic (Stams 1994). Combination of propionate degradation (1.1) with hydrogenotrophic (1.3) and aceticlastic (1.2) methane production gives the Gibbs free energy of about -56 kJ/mol propionate (at standard conditions) which means 1 ATP is available for all associated organisms. That means living at the absolute energetical minimum that is thermodynamicly possible. Müller et al. (2010)

titled this phenomenon "the smallest quantum of energy in biology", as about 1/3 ATP is net available for propionate degrader at the end of its conversion.

In such severe conditions the participants must organize each other in a manner giving them the possibly to vanquish the growth inconveniencies. As they have to cooperate, the natural consequence is living in groups, where the distances between organisms are relatively low. That should enhance the interspecies hydrogen/ formate transfer, which is of utmost importance for the cooperation between propionate degraders and methane producers. This transfer occurs through diffusion and can be expressed by Fick's diffusion law. Both, hydrogen and formate can diffuse between organisms in a liquid media, but as the hydrogen interspecies transfer is better understood, further considerations will be mostly based on this electron carrier, as expressed in (1.8).

$$J_{H_2} = A_{PA} \cdot D_{H_2} \frac{C_{PA} - C_{H_2}}{d_{PAH_2}} \tag{1.8}$$

J_{H_2} - hydrogen flux between hydrogen producer and hydrogen consumer

A_{PA} - surface area of propionate degrader (hydrogen producer)

D_{H_2} - hydrogen diffusion constant in water

C_{PA} - hydrogen concentration outside hydrogen producer

C_{H_2} - hydrogen concentration outside hydrogen consumer

d_{PAH_2} - distance between hydrogen producer and hydrogen consumer

Small interspecies distances can be maintained by growth in flocs or aggregates or by increasing the population density of hydrogen consumers, which lead to increased propionate-degradation rates (Dwyer et al. 1988; Thiele et al. 1988; Schmidt and Ahring 1995; Imachi et al. 2000; Ishii et al. 2005). As diffusion is the major process of hydrogen transport between microbial species (McCarthy and Smith 1986), it appears from Fick's diffusion law (1.8) that the hydrogen flux depends inter alia on the distance between H_2-producing and H_2-consuming organisms. The same consideration would be valid for formate flux calculations (Stams 1994). The standard redox potential of the H^+/ H_2 couple is -414 mV and is almost the same as for the CO_2/ formate couple (-420 mV) making both electron transfer reactions equally important in syntrophic interactions (Schink 1997). Formate formation and degradation governed by the hydrogen partial pressure is of importance for electron transfer in syntophic cultures. Interspecies distances for an optimal electron transfer have until now only been calculated theoretically. Computations with the above mentioned diffusion law revealed interspecies distances of syntrophic propionate oxidizers and methanogens of less than 3 µm (Schmidt and Ahring 1995), 2 µm (Ishii et al. 2005), 1 µm (Stams et al. 1989; Imachi et al. 2000) or even only 0.08 µm (Shi-yi and Jian 1992). This distance is schematically presented in Figure 3 (denoted as "d"). The cell dimensions of described propionate degraders (denoted as "A" in Figure 3) are within 0.6–3.0 µm x 1.0–11.0 µm (Wallrabenstein et al. 1994, Wallrabenstein et al. 1995, Nilsen et al. 1996, Harmsen et al. 1998, de Bok et al. 2001, Imachi et al. 2002, de Bok et al. 2005, Chen et al. 2005) and the methanogenic partners (denoted as "B" in Figure 3) on the example of *Methanospirillum hungatei* may reach 0.5–7.4 µm x 15– several hundreds µm (Ferry et al. 1974).

Above description implies that the distance between partners in the syntrophic relationship plays a significant role in hydrogen interspecies transfer. The research done before confirms that increased propionate degradation rates had a direct connection to shortening the span between organisms by inducing precipitation or adding methanogens to the culture (Dwyer et al. 1988, Schmidt and Ahring 1995). The positioning of a cell of a hydrogen producing/ propionate degrading organism and a cell of a hydrogen consuming/ methane producing organism at a certain distance between each other is illustrated in Figure 1.3.

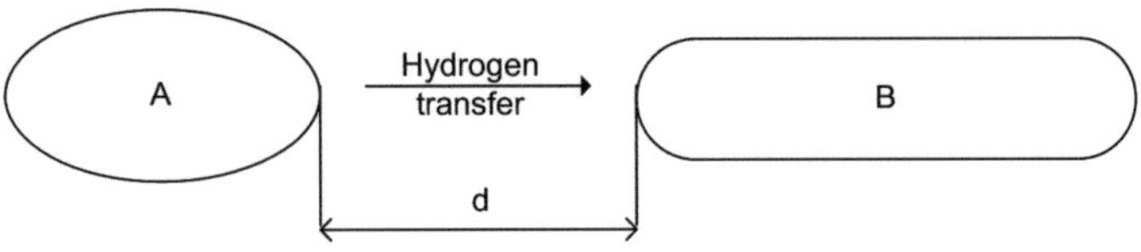

Fig. 1.3: Hydrogen transfer between propionic acid degrading bacterial cell (A) and methane producing bacterial cell (B) positioned at a certain distance (d) from each other.

The granulation during organic waste treatment in UASB reactors has been pointed out as important in terms of its function. It gathers the microorganisms participating in biomethanisation and is considered as a unit enhancing syntrophic degradation of metabolites. Their structure is an object of several studies and has been analyzed by e.g. Sekiguchi et al. (1998), Schmidt and Ahring (1999) or Kim et al. (2012). It was proposed that a granule has a layered structure, consisting of an aceticlastic methanogens' core surrounded by acetogenic (hydrogen/ formate producers) and hydrogen/ formate consuming methane bacteria (McCarty and Smith 1986, Liu et al. 2003) as shown in Figure 1.4. Such structure supports optimal propionic acid degradation, as it allows the reduction of the spatial limitation for interspecies electron transfer. However, the granule structure may differ according to the conditions in the reactor, and a shifting of layers of different microorganisms may be observed (Kim et al. 2012).

Methanoseta concilii is thought to be the crucial initiator of granulation as it forms the nucleation centre and was found remaining mostly in the granule center during the operational time of UASB reactors (Zheng et al. 2006). Liu et al. (2003) described this as one of the several models of granulation initiation possibilities. The inert nucleus model basing on microbial attachment to zeolite, the multi-valence positive ion-bonding model using electrostatic interactions between positively charged ions added to sludge and negatively charged bacteria or extracellular polymer bonding model are other examples of this phenomenon (Liu et al. 2003).

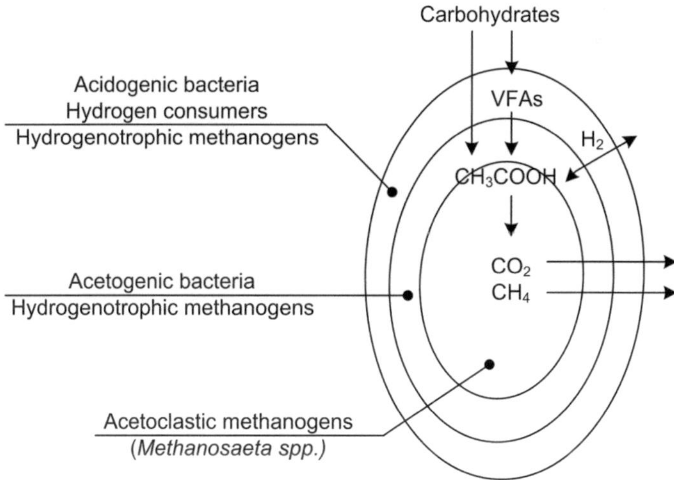

Fig. 1.4: Layered structure of an upflow anaerobic sludge blanket reactor granule according to Liu et al. (2003).

Propionate degrading bacteria enhance formation of granules in UASB reactors, by "coating" the *M. concilli* nucleus as reported by Zheng et al. (2006). The size of granules was increasing faster in the UASB reactor fed with propionate, than in the parallel one running on feed without propionate addition. The first one contained a larger number of *Syntophobacter sp.* that were a constituent of juxtaposed archaea and bacteria at the subsurface of a granule, like presented in Figure 1.4 . Harmsen et al. (1996) analyzed UASB granules from a reactor fed with propionate or propionate and sulfate. The samples from the first one had granules with microcolonies consisting of methanogens and microcolonies of syntrophic propionate degrading bacteria (SYN7) intertwined with hydrogenotrophic methanogens' chains. During another study, Tatara et al. (2008) found *Pelotomaculum-* related clones in the biofilm of a thermophilic down-flow anaerobic packed-bed reactor, where propionate degradation occurred at high rate (Table 5.3) thanks to its close positioning to *Methanothermobacter*-related clones attached to the packed-bed. Stams et al. (1989) showed a transmission electron micrograph of organisms from a granular sludge grown on propionate, that *Syntrophobacter*-like cells were surrounded by *Methanobrevibacter sp.*, proving that they must be positioned in close vicinity.

Above results demonstrate that propionate degrading bacteria live in a close neighborhood of hydrogen/ formate utilizing methanogens, presumably to enhance the hydrogen/ formate flux by minimizing the distances between them. However the exact model of the "gathering" process specifically between these two bacterial groups is not clear. According to Schink and Stams (2006), the aggregation of defined co-cultures strongly depends on the chosen participants, e.g. *Methanospirillum hungatei* is not the best partner for the formation of stable granules, and that sets certain boundaries for the research in this area. They proposed that random mixing of partner organisms within a granule would be more beneficial than formation of microcolonies' combination ("nests"), like presented in Figure 1.5. Nevertheless, within aggregates' growth in time, the "nests" structure is more probable, unless internal mixing of microorganisms could occur.

One of the curiosities in the findings on co-aggregation possibilities and interspecies cooperation on propionate degradation is the finding of Gorby et al. (2006) that Pelotomaculum thermopropionicum can form electrically conductive pilus-like filaments to coaggreagate with methanogens and that the electron transfer between these bacteria proceedes also through filaments called bacterial nanowires.

1.4.3 Fluorescence in-situ hybridization as one of the methods for microbial identification and population change analysis

Microbial community analysis in anaerobic digestion of biowaste is a challenge, considering the variety of participating microorganisms during each process step. Modern molecular techniques in biology allow however the

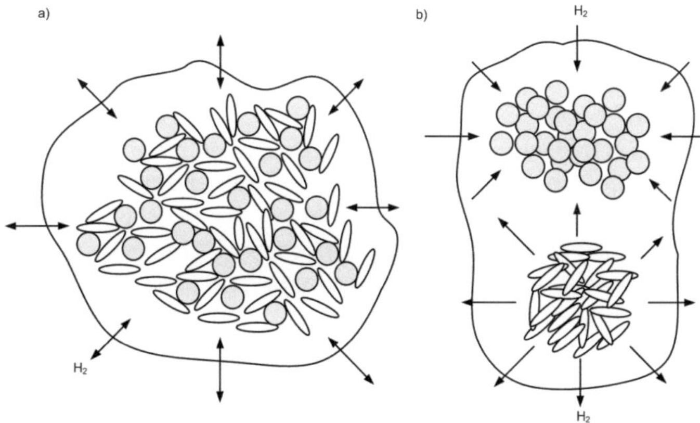

Fig. 1.5: Schematic representation of hydrogen exchange in differently arranged anaerobic bacterial aggregates: a) homogeneous distribution of hydrogen producers and hydrogen consumers, b) separated nests formation of hydrogen producers and hydrogen consumers (Schink and Stams 2006).

insight into microbial population structure and dynamics. The hybridization of 16S rRNA-based oligonucleotide probes like dot blotting (Angenent et al. 2002) and FISH (Harmsen et al. 1996, Ariesyady et al. 2007, Park et al. 2009), 16S rRNA gene clone libraries analysis (Fernandez et al. 2000), the 16S rRNA genes analysis by polymerase chain reaction single-strand conformation polymorphism (PCR-SSCP) (Zumstein et al. 2000, Chachkhiani et al. 2004), analysis of polar lipid fatty acid (Oude Elferink et al. 1998), immunohistochemical analysis (Schmidt and Ahring 1999) and rRNA based stable-isotope probing (SIP) (Leueders et al. 2004, Hatamoto et al. 2007) are applied approaches in abundance and functions examination of microorganisms in biogas reactors.

Fluorescence *in-situ* hybridization has a function of establishing the quantitative microbial diversity and, mostly in combination with confocal laser scanning microscopy (CLSM), the spatial distribution of several community groups within aggregates/ flocs/ granules (if formed). Hybrydization of short DNA probes mono-labeled with a fluorescent dye to ribosomal RNA of the whole cells treated previously with a permeabilizing agent is the principle of this technique (Nielsen et al. 2009). The possibility of designing probes that are specifically targeting a wide phylogenetic group (e.g. *Eubacteria*) or only for organisms at the species level (e.g. *Syntrophobacter fumaroxidans*) is beneficial. Furthermore, the method can be applied to obtain information of a microbial composition directly (*in-situ*) in a sample without changing cell morphology, what is crucial for organisms' spatial organization recognition and their quantification. Daims (2009) uses the direct visual feedback of analyzed sample and "count what you see" approach as the key advantage over PCR-based techniques that are more sensitive to methodical errors. However, such problems like insufficient penetration of a probe into bacteria, low ribosome content and its inaccessibility or sample autofluorescence are the obstacles that have to be taken into account (Liehr 2009, Nielsen et al. 2009). Additionally, FISH is considered to be a laborious approach with no automatisation potential.

The technique of FISH is widely applied in research on syntrophic bacteria capable to degrade propionic acid, especially in terms of their spatial distribution within granules. Kim et al. (2012) published recently the results on the microbial distribution in such structures formed in mesophilic two stages UASB reactors treating purified terephthalic acid wastewater (where two subunits were identical) with FISH application combined with CLSM. Granules formed in each stage differently organized. First stage acclimatized granules were formed under the conditions that were favorable for methanogens and methanogenic bacteria were situated at the exterior layer, whereas the second stage granules showed a microbial distribution typical for acetogenic-favorable conditions, where the outer layer consisted of acetogenic microorganisms. Zheng et al. (2006) examined the population dynamics during the granulation process and the granule structure itself in UASB reactors with FISH as well. Harmsen et al. (1995) verified the specificity of MPOB1 and KOP1 oligonucleotides by means of in-situ hybridization of targeted organisms with rhodamine labeled probes. The same authors analyzed in 1996 the dynamics of propionate-oxidiziers population in anaerobic granular sludge under methanogenic and sulfidogenic conditions using 16S rRNA-based oligonucleotides. Imachi et al. (2000) investigated the abundance and spatial distribution of these degraders in the thermophilic sludge granules by FISH with CLSM application. The ratio of aceticlastic and hydrogenotrophic methanogens was determined with this technique in a thermophilic dry (containing from 20 % to 25 % total solids) anaerobic reactor treating the organic fraction of municipal solid wastes in order to find the correlation between VFAs degradation and population changes of methanogens (Montero et al. 2010). It was found that butyrate consumption was related to increasing numbers of hydrogenotrophic methanogens during the start-up period and to aceticlastic methanogens during reactor stabilization. FISH was also proved to be a good method, combined with other reactor parameters, for reactor efficiency control (Fall 2002).

The combination of FISH with microaudiography (FISH-MAR) allows the determination of radiolabeled substrate assimilation and the function of microorganisms (Chong et al. 2012). FISH-MAR was applied by Ariesyady et al. (2007) for analysis of propionate degrading bacteria, where an uncultured *Smithella* long rod phylogenetic group was identified by a positive signal of propionate assimilation in microaudiography, but no response after sample hybridization with *Smithella* short rod specific oligonucleotide probe. That resulted with new oligonucleotide probe designing- *Smithella* long rod specific. The FISH-MAR image examples of anaerobic digester sludge are presented in Figure 1.6.

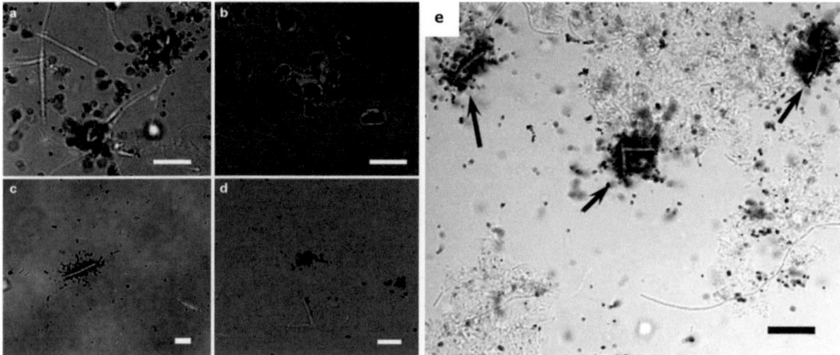

Fig. 1.6: The FISH-MAR (microautoradiography) images of propionate degraders in anaerobic digester sludge taken by Ariesyady et al. (2007), a) TRITC labeled Smi SR 354, b) TRICT labeled Synbac 824, c) FITC labeled Eub 338, d) TRICT labeled Eub338, e) FITC labeled Smi LR 150; bar represents 5 µm (a-d) and 10 µm (e).

Catalyzed reporter deposition (CARD)-FISH is a modified method using horseradish peroxidase (HRP)-labeled probes and amplification of one tyramide signal and is recommended for samples containing cells with low ribosome contents. It is a modification of a standard FISH technique where the detection limit in environmental samples was diminished from 1400 to 30–60 target molecules (Wagner and Haider 2012).

2 Aim of the work

Propionic acid accumulation during anaerobic treatment of biowaste is a common reason of digesters upset and biogas production stagnation. The stagnation period may last up to a few months until the disrupted microbial flora regenerates and the biogas yield returns to the optimal level (about 65 % methane content). The main reasons for this are however not yet known, and the microorganisms participating in the process are not fully described. The deeper insight into the relationship between bacteria responsible for propionate degradation is needed to understand their behavior and to design a plan for preventing or healing approaches. The goal of this thesis is to describe propionate degradation during anaerobic biowaste digestion in different digester regimes through:

- development of an analytical method for propionic acid metabolic pathway determination by combining headspace (HS), gas chromatography (GC) and mass spectrometry (MS);

- application of the developed method for predominant pathway identification in samples containing microorganisms of different origin: an industrial scale CSTR reactor treating biowaste, Indonesian lab-scale reactors treating market waste and enrichment cultures on propionate from these digesters;

- preparation of enrichment cultures on propionic acid with addition of several co substrates in order to isolate effectively degrading cultures and microbial description of syntrophic bacteria with the use of fluorescence *in-situ* hybridization (FISH);

- investigation of the ability of propionate degradation by the organisms from different habitats- mesophilic, thermophilic, halophilic, and comparing obtained degradation rates;

- examination of the influence of interspecies distances on optimal hydrogen transfer and observation of microbial aggregate formation.

The combination of applied modern analytical chemistry, standard microbiological approach and molecular biology for describing and explaining biotechnological problem of propionic acid accumulation during anaerobic biowaste degradation should allow making a meaningful step towards a better understanding of the biomethanisation process. Description of bacteria by means of the digestion pathway is essential for their identification, just like observing species morphology under the microscope. Furthermore, the examination and comparison of organisms' behavior under different conditions (and of different origin) should point out the microorganisms characterized by the fastest degradation rates and thus be useful in finding optimal parameters for undisrupted methane production by eliminating the problem of propionic acid accumulation. FISH method applied for microbial identification should extend the results obtained during the examination of metabolic pathway and allow denoting the most abundant species responsible for propionate degradation in different habitats. Additional information about the spatial distribution of methanogens and propionate degraders may be helpful in selecting and/ or designing anaerobic reactors enhancing their growth by e.g. supplying suitable conditions for optimal interspecies hydrogen transfer.

3 Materials and methods

3.1 Microorganisms preparation

3.1.1 Cultivation of mixed cultures

Microorganisms from different habitats were collected for the purposes of the research on anaerobic propionic acid degradation. These were:

- thermophilic full-scale fermentation plant- one stage, one phase dry fermentation (Leonberg, Germany);

- mesophilic full-scale fermentation plant- one stage, one phase wet fermentation (Durlach, Germany), presented in Figure 3.1;

- mesophilic lab-scale fermentation plant- one stage, two phase wet and dry fermentation (Institute of Technology Bandung (ITB), Indonesia), presented in Figure 3.2;

- mesophilic lab-scale fermentation plant- two stages, one phase wet fermentation (Indonesian Institute of Sciences (LIPI), Bandung, Indonesia), presented in Figure 3.3;

- halophilic sediments from the North Sea, incubated at room temperature.

For the purposes of the determination of propionic acid oxidation pathway, microorganisms were prepared as enrichment cultures on propionic acid as a sole carbon source for their growth. At first, if there were any VFAs present in the sample, the pH of the bacterial suspension was adjusted to 7, and incubated until all acids were degraded. Afterwards, propionic acid was added, and its degradation rate was monitored. Then, the organisms were transferred into mineral medium and propionic acid was added (biomass was usually introduced at a level of 10 % of medium volume). Contents of used mineral broth are listed in Table 3.1.

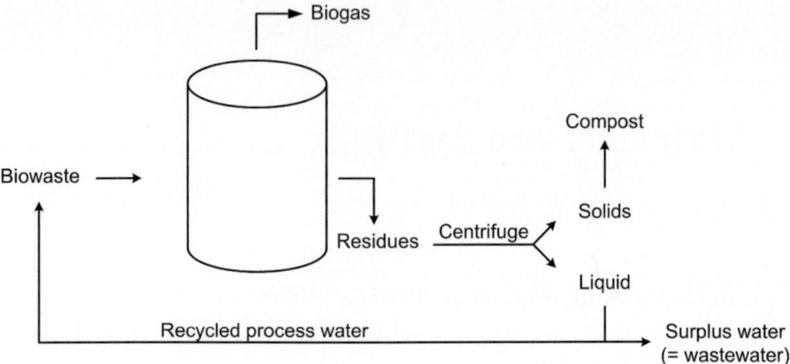

Fig. 3.1: Mesophilic full-scale fermentation plant Durlach, Germany.

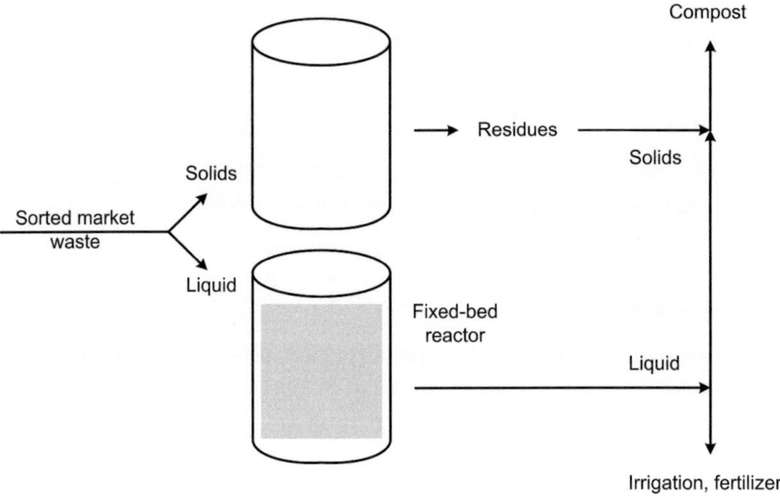

Fig. 3.2: Mesophilic one stage, two phases laboratory-scale fermentation plant Bandung, Indonesia.

NaOH, Resazurin and H_2O were heated until their boiling point was reached to make the solution anaerobic, and then the Na_2S was dissolved. The mixture was poured into Schott glass flasks, closed with a rubber stopper and secured with a plastic screw cap. The gas atmosphere was exchanged to N_2 and the solution was prepared for further use.

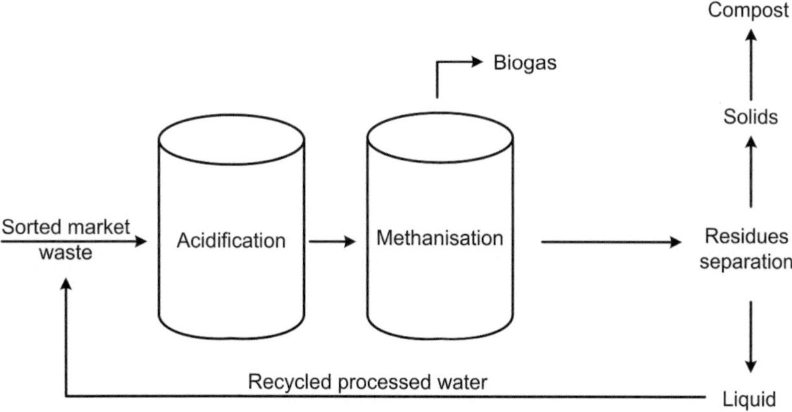

Fig. 3.3: Mesophilic two stages, one phase laboratory-scale fermentation plant Bandung, Indonesia.

The medium was specially prepared to make it anaerobic. At first, all the agents listed in Table 3.1, excluding $NaHCO_3$ and reducing agent, were dissolved in 986.4 ml bidestilled water and mixed. The pH was measured with a glass electrode and adjusted to 7.4 with 0.1 N HCl. Heating of the solution until it boiled was the next step. Boiling liquid was poured into the Schott flask and secured with a rubber stopper and a plastic screw cap. The atmosphere was exchanged with N_2 at the gas station, taking the advantage of the elevated temperature and "boiling" it under low pressure formed during evacuation. That allowed maximal oxygen removal from the liquid. After two gas exchange cycles the reducing agent was introduced into the medium with a syringe. The next step was conducted in the anaerobic chamber, where, after the oxygen level dropped sufficiently (down to 1 ppm), $NaHCO_3$ was dissolved in the solution. The medium was then distributed into serum bottles of smaller volume- according to needs. The smaller serum bottles were then closed with a rubber stopper, taken out of the anaerobic chamber to the gas station, where the atmosphere was again exchanged for nitrogen. The gas exchange was necessary, as the atmosphere in the anaerobic chamber consisted of CO_2 and H_2 (95:5 %, v/ v) and hydrogen would affect propionic acid degraders growth at low levels already. Prepared medium was autoclaved at 121 °C for 20 min, and used for bacterial growth experiments according to needs.

Table 3.1: Constituents of the mineral medium for propionic acid degraders.

Substance/ solution	Amount/ volume	Purity	Reference
NH_4Cl	$0.5\,gl^{-1}$	p.a.	-
K_2HPO_4	$10\,mmoll^{-1}$ $(= 1.74\,gl^{-1})$	p.a.	-
$CaCl_2 / MgCl_2$ [1]	$1\,mll^{-1}$ $(60\,mgl^{-1}$ each)	p.a.	-
$NaHCO_3$	$3.8\,gl^{-1}$	p.a.	-
SL 10 [2]	1 ml/l	-	Tschech and Pfennig, 1984
Se, WO_4 [3]	$0.5\,mll^{-1}$	-	Tschech and Pfennig, 1984
Wolfes' vitamins [4] (10 x concentrated)	$1\,mll^{-1}$	-	Balch et al. 1979
Resazurin (0.1 %)	$0.1\,mll^{-1}$	p.a.	-
Reducing agent [5]	$10\,mll^{-1}$	-	-

[1] $CaCl_2/ MgCl_2$ stock solution:
- $CaCl_2$ x 2 H_2O (6 g)
- $MgCl_2$ x 6 H_2O (6 g)
- H_2O bidest. (100 ml)

[2] SL-10 trace element solution:
- HCl $(25\%; 7.7\,moll^{-1})$ (10 ml)
- $FeCl_2$ x 4 H_2O (1500 mg)
- $ZnCl_2$ (70 mg)
- $MnCl_2$ x 4 H_2O (100 mg)
- H_3BO_3 (6 mg)
- $CoCl_2$ x 6 H_2O (190 mg)
- $CuCl_2$ x 2 H_2O (toxic) (2 mg)
- $NiCl_2$ x 6 H_2O (toxic) (4 mg)
- Na_2MoO_4 x 2 H_2O (36 mg)
- H_2O bidest. (990 ml)

$FeCl_2$ was dissolved in HCl and than diluted with water, next all salts were added and the solution was filled up to 1000 ml.

[3] Selenite/ tungstate solution
- NaOH (500 mg)
- Na_2SeO_3 x 5 H_2O (3 mg)
- Na_2WO_4 x 2 H_2O (4 mg)
- H_2O bidest. (1000 ml)

[4] Wolfe's vitamin solution
- D (+) Biotin (2 mg)
- Folic acid (2 mg)
- Pyridoxamindihydrochloride (10 mg)
- Thiaminiumdichloride (5 mg)
- Riboflavin (5 mg)
- Niacin (5 mg)
- Ca-(D+)-Pantothenate (5 mg)
- Cyanocobalamin (1 mg)
- p-Aminobenzoic acid (5 mg)
- DL-α-Lipoic acid (5 mg)
- H_2O bidest. (1000 ml)

Prepared as 10 times concentrated solution (all amounts 10 times multiplied) and stored at 4 °C in the darkness.

[5] Reducing agent solution
- NaOH $(1\,moll^{-1})$ (10 ml)
- Resazurin (0.1 %) (0.1 ml)
- Na_2S x 9 H_2O (1200 mg)
- H_2O bidest. (90 ml)

The only modification made for halophilic propionate degraders, was an addition of $16\,gl^{-1}$ NaCl. The atmosphere was exchanged to $N_2/\ CO_2$ (80:20 %, v/ v).

3.1.2 Bacterial enrichments

For bacterial isolation purposes, propionic acid was used in the liquid form and was diluted from 99 % propionic acid (p.a; Fluka, Germany) in pre-boiled distilled water up to 1 mol l^{-1} concentration, closed immediately with a rubber stopper, and autoclaved. Before autoclaving, the the atmosphere was exchanged with N_2. Similarly additional co-substrates like pyruvate, lactate, fumarate and sulfate were prepared. The only exception was atmosphere exchange performed while the solution had elevated temperature, as these were non-volatile chemicals.

Medium in solidified form for the closed roll tube isolation method and for Petri Dishes was prepared according to Table 3.1 with addition of 20 g l^{-1} agar. After autoclaving and cooling down to appropriate temperature (40 °C, the roll tubes were inoculated with 10 % (v/ v) suspended microorganisms and incubated at appropriate conditions. Colonies were picked in the anaerobic chamber with a sterile needle, prior their transfer into the liquid medium.

Liquid medium for propionic acid degraders isolation was distributed into serum bottles of 120 ml volume, filled with 40 ml liquid medium, leaving 70 ml gaseous volume. The medium was supplied with propionic acid or propionic acid and a co-substrate, depending on the culture type. The glasses were closed with a rubber stopper and secured with an aluminum cap. After biomass enrichment (usually when 20 mmol l^{-1} propionic acid was consumed; after about 40 days incubation), the transfer of microorganisms into a fresh medium was done (10 % volume). Organisms were fed with 5 mmol l^{-1} propionic acid portions with a syringe flushed prior substrate addition with sterile nitrogen (5.0) captured in a serum bottle.

Dilution series were prepared in 25 ml volume Rollrand tubes, filled with 10 ml medium. The inoculation principle was transferring of a 10 % (v/ v) inoculums according to Figure 3.4.

Pure cultures

Pure cultures of *Syntrophobacter fumaroxidans* (DSM-10017) and of *Smithella propionica* (DSM-16934) were purchased from German Collection of Microorganisms and Cell Cultures (DSMZ) and cultivated in anaerobic media 684 and 1030, respectively, according to DSMZ. *Enterococci* were kindly provided by Dr. Schmidt (KIT, IBA). These bacteria were used as control organisms for FISH and spore staining.

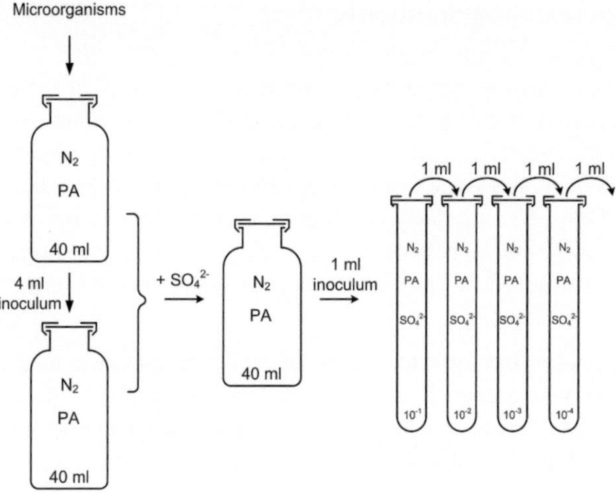

Fig. 3.4: Example of experimental principle for propionic acid degraders' isolation, here with sulfate addition as a co-substrate; PA = propionic acid; N_2 = nitrogen gas phase; 40 ml = medium volume; SO_4^{2-} = e^--acceptor.

3.2 Volatile fatty acids analysis with HS-GC-MS and determination of metabolic pathway of anaerobic oxidation of propionic acid

Prior to analysis, samples were transferred into glass vials closed with a septum and secured with an aluminum cap and incubated in a carrousel immersed silicon-oil bath of the HS device. After appropriate time the gas-analyte vapor mixture was collected with HS needle into the sampling loop and then transferred through the heated transfer line into the GC, where analytes were separated. Separated compounds entered the MS through the heated transfer line and the detector signal was sent to the processing unit (personal computer, PC). The headspace sampling unit was programmed through a separate control station. Parameters for GC and MS were controlled by PC. Apparatus arrangement is shown in Figure 3.5.

The GC was equipped with a 30 m CP-Wax FFAP column of 0.25 mm internal diameter (ID). The column stationary phase was a polyethylene glycol polymer modified with a nitroterephthalic acid (Figure 3.6), specific for free volatile fatty acid analysis and their methyl esters. As carbon dioxide could not

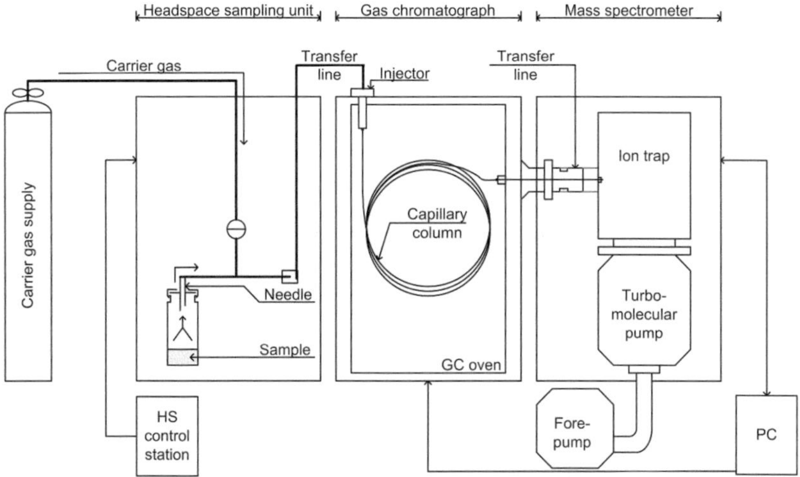

Fig. 3.5: Schematic representation of HS-GC-MS system for VFAs analysis.

be detected with the column used, the research was focused on identification of acetic acid fragmentation. For differentiation between the two metabolic pathways, the amounts of produced acetate were determined and the ^{13}C-labeling at methyl- or carboxyl moiety was investigated: If the methyl malonyl CoA pathway was active, oxidation of 1-^{13}C-labeled propionate should result in the production of 1 mol non-labeled acetic acid (with a molecular ion of 60 m/ z at the MS spectrum) and 1 mol ^{13}C-labeled carbon dioxide Figure (3.7). If the C-6-dismutation pathway was active, 3 mol of acetic acid should be generated from 2 mol of 1-^{13}C-labeled propionate, where 1 mol is non-labeled, 1 mol is ^{13}C-methyl- and 1 mol is ^{13}C-carboxyl-labeled, both with 61 m/z molecular ion in the MS spectrum (Figure 3.8). Detailed mass balance for possible pathways of propionic acid degradation and possible metabolites is included in Table 3.2.

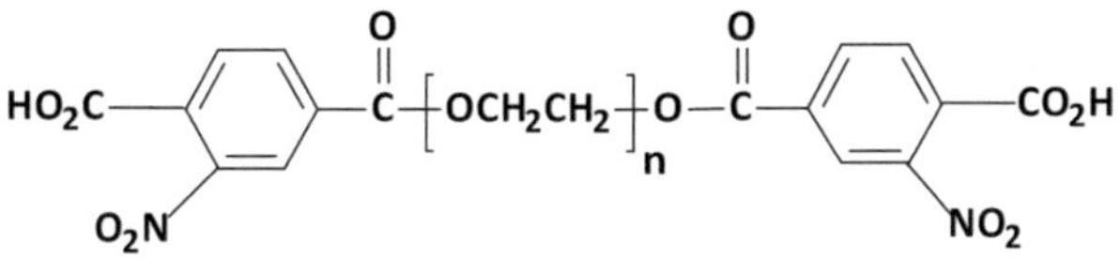

Fig. 3.6: Polyethylene glycol modified with nitroterephthalic acid.

31

The substrate used, 1-^{13}C-propionic acid should give in the case of the methyl-malonyl-CoA pathway non-labeled acetic acid with a molecular mass of $60\,\mathrm{g\,mol^{-1}}$ giving 60 m/z molecular ion on the mass spectrum and three different forms of acetic acid if the C-6-dismutation pathway (Figure 3.8) was active. Fragmentation ions for all possible metabolites and the substrates used, as well as detailed mass balance are represented in Table 3.2. The main ionization products are also represented in the mass spectra from NIST library (Figure 3.9).

Table 3.2: Mass balance and main fragmentation ions for substrate and metabolites

Metabolic pathway	Substrate [mol]	Products [mol]				
	$CH_3CH_2^{13}COOH$	$^{13}CH_3COOH$	$CH_3^{13}COOH$	CH_3COOH	$^{13}CO_2$	CO_2
Methyl-malonyl-CoA	1	–	–	1	1	–
C-6 dis-mutation	2	1	1	1	–	–
Fragmen-tation ions	28, 46 58, 75	16, 44 45, 61	15, 44 46, 61	15, 43 45, 60	13, 16 29, 45	12, 16 28, 44

3.3 Non-volatile intermediate metabolites analysis

Method for quantitative and qualitative of non-volatile acids analysis was applied for co-substrates concentration monitoring during microbial growth. Dani 3950- HS Varian 410-GC- Varian 210-MS was used at operational parameters listed in Table 3.3. Sample preparation prior analysis started with centrifugation at 13000 rpm for 2 min in 1.5 ml Eppendorf caps. 20 µm of supernatant together with 10 µm 50 % (w/ v) $NaHSO_4 \cdot H_2O$ and 10 µm MeOH ($\geqslant$ 99.5 %; HPLC purity) were transferred into 20 ml glass vials and closed with a silicone septum and an aluminium cap. Before analysis the analytes in the acidified sample underwent conversion into volatile methyl esters, as described previously by Heitefuss et al. (1990). For this purpose, temperature and incubation time of the sample in the silicon oil bath of HS were set as listed in Table 3.3. Temperature was set to 120 °C and the time was extended to 20 min, so the esterification of non-volatile compounds could occur and their measurement in the form of methyl esters was possible. Comparison of

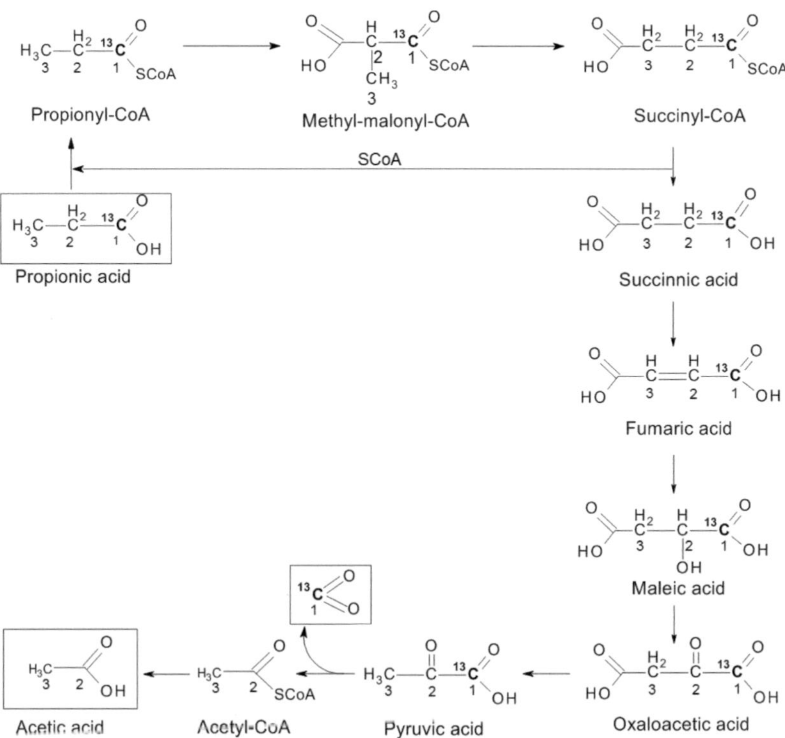

Fig. 3.7: The distribution of ^{13}C during the methyl-malonyl-CoA pathway of propionic acid oxidation (pathway according to Plugge et al. 1993).

the boiling points for compounds and their methyl esters are shown in Table 3.4.

Table 3.3: Parameters set for non-voaltile acids anaysis for HS-GC-MS system.

HS - DANI 3950	GC - Varian 431C	MS - Varian 210
Liquid volume: 40 μl	Injector temperature: 250 °C	Ionization mode: EI
Bath temperature: 120 °C	Split ratio: 1:10	Ionization energy: 70 eV
Incubation time: 20 min	Temperature program:	Trap temperature: 180 °C
Gas sampling time: 10 s	70 °C $\xrightarrow{6\,°C/min}$ 120 °C $\longrightarrow$	Axial modulation: 2.2 V
Gas sample volume: 1 ml	$\xrightarrow{30\,°C/min}$ 200 °C	
	Carrier gas: He (5.0)	
	Carrier gas flow: 1.0 ml/min	
	Total analysis time: 14 min	

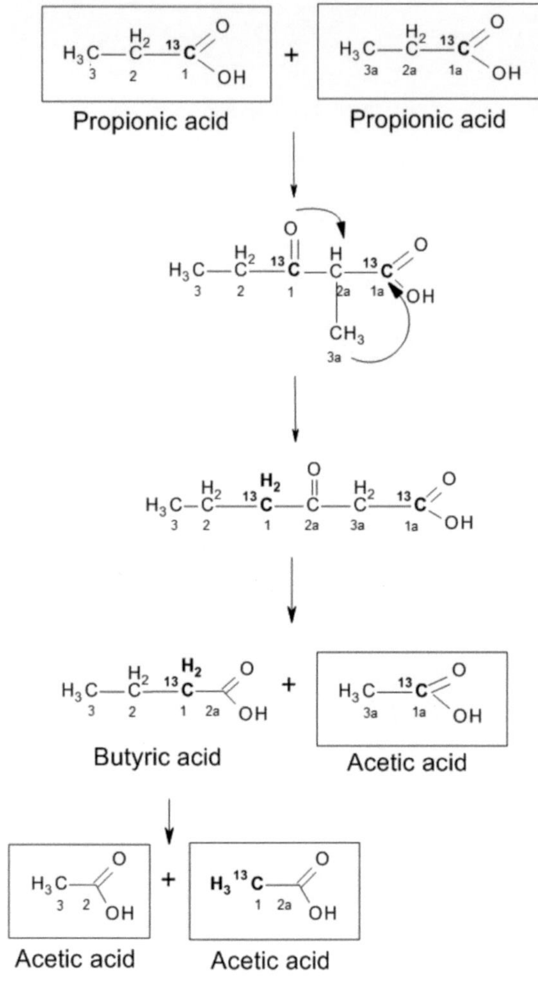

Fig. 3.8: The distribution of ^{13}C during the C-6-dismutation pathway of propionic acid oxidation (pathway according to de Bok et al. 2001).

Following non-volatile compounds were measured: pyruvic acid, lactic acid, fumaric acid and succinic acid. After chromatographic separation they were eluting in the same order with retention times of 3.475 ± 0.009, 4.781 ± 0.004, 9.078 ± 0.008 and 10.020 ± 0.010 min respectively. Identification of obtained mass spectra of the methylated compounds gave the following compounds

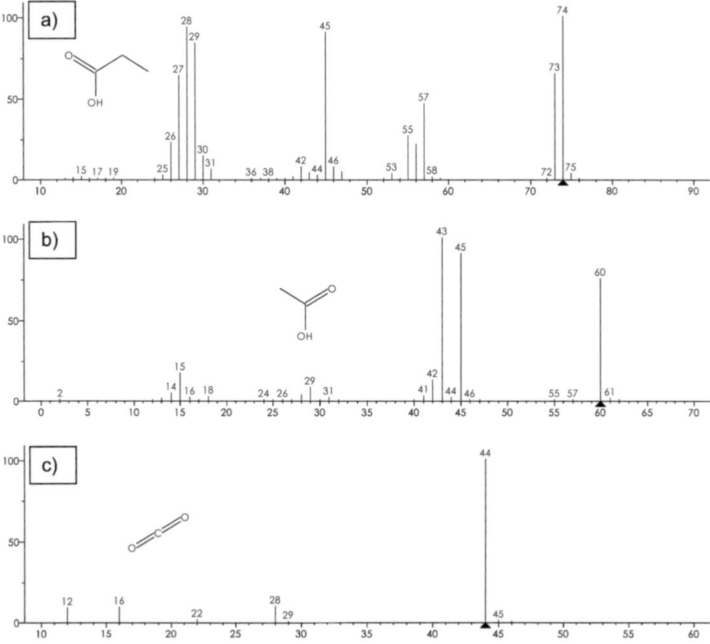

Fig. 3.9: Mass spectra according to NIST library for a) propionic acid, b) acetic acid, c) carbon dioxide; ▲ denotes molecular ion.

according to NIST spectra library: methyl, 2-oxopropionate; propanoic acid, 2- hydroxy-, methyl ester; 2- butenedioic acid, dimethyl ester; and butanedioic acid, dimethhyl ester. Chromatographic separation is presented in Figure 3.10. Calibration plots for non-volatile compounds were prepared with the use of fumaric acid disodium salt (Merck-Schuchardt, Germany), 85 % (w/w) DL- lactic acid, pyruvic acid sodium salt (Carl Roth GmbH, Karlsruhe, Germany) and succinic acid at a concentration of $1\,mmol\,l^{-1}$, $2\,mmol\,l^{-1}$, $3\,mmol\,l^{-1}$, $4\,mmol\,l^{-1}$ and $5\,mmol\,l^{-1}$, respectively. Regression coefficients for compounds were in the range from 0.993 to 0.955, giving sufficient correlation between measurements for reliable analysis of analytes.

Table 3.4: Non-volatile metabolites characteristic and their methyl esters.

Compound	Synonym	Boiling point	CAS number	Methyl ester	Boiling point	CAS number
Pyruvic acid	2-oxo-propanoic acid	165 °C	127-17-3	Methyl-2-oxopropionate	134 °C to 137 °C	600-22-6
LD-Lactic acid	2-hydroxy-propoanoic acid	122 °C	50-21-5	2-hydroxy-propanoic acid methyl ester	35 °C	547-64-8
Fumaric acid	Butenedioic acid	295 °C to 300 °C	110-17-8	(E)-2-butenedioic acid dimethyl ester	192 °C to 193 °C	624-49-7
Succinic acid	Butanedioic acid, amber acid	185 °C to 190 °C	110-15-6	Methyl butanedioate	200 °C	106-65-0

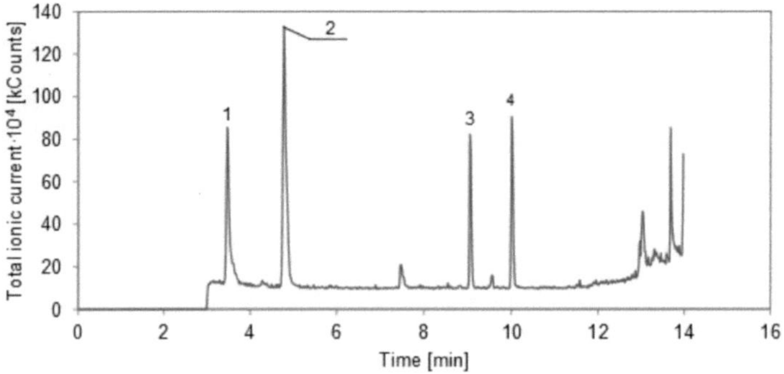

Fig. 3.10: Chromatographic separation of methylated compounds 1) pyruvic acid methyl ester, 2) lactic acid methyl ester, 3) fumaric acid dimethyl ester, 4) succinic acid dimethyl ester.

3.4 Gas chromatographic determination of gases

Hydrogen and methane were analyzed with a Gas Chromatograph Chrompak CP 9001 after separation with a capillary column CarboPlot P7 WLD FS 25 x 0.53 m (Chrompack, Netherlands) by a micro volume thermal conductivity detector (TCD). Nitrogen carrier gas flow was set to 30 ml/min. Temperatures of the injector, the oven with the column and of the detector were 250 °C, 110 °C and 220 °C, respectively. Standard mixtures of gases were prepared from 5.0 purity gases, mixed appropriately to give the desired concentration. Concentration was measured as a volume percentage occupied by considered gas in a sample. They were prepared in a serum bottle secured with a rubber stopper and an aluminium cap. Injections into the column were performed with a 0.5 ml Pressure-Lok syringe (VIVI Precision sampling, USA). The injection volume was 0.1 ml. Calibration curves for hydrogen and methane are represented in Figure 3.11.

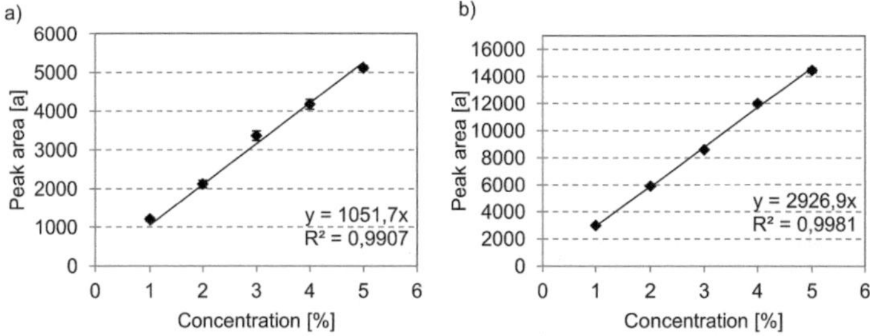

Fig. 3.11: Calibration curve for a) methane and b) hydrogen.

3.5 Ion chromatography for sulfate determination

Sulfate anions were analyzed with an ICS-90 Ion Chromatograph from Dionex, equipped with an Ion-Pac AS9-HC column and a suppressed conductivity detector. $90\,mmol\,l^{-1}$ Na_2CO_3 solution was used as an eluent (flow rate: $1\,ml/min$) and $36\,mmol\,l^{-1}$ sulfuric acid as a regeneration solution. Before analysis samples were centrifuged at 15000 rpm for 5 min in 1.5 ml Eppendorf caps. Clarified supernatant was then diluted with ultrapure Milli-Q water at a ratio 1:10 and injected into the system (1 ml sample volume).

Sulfate sodium salt was used to prepare standard solutions for a calibration curve (Figure 3.12). The salt was diluted in ultra pure MilliQ water at concentrations $1\,mmol\,l^{-1}$, $2\,mmol\,l^{-1}$, $3\,mmol\,l^{-1}$, $4\,mmol\,l^{-1}$ and $5\,mmol\,l^{-1}$.

To check if the test was reliable, a comparison with spectrophotometric approach was done. Sulfate in the form of sodium salt was diluted in MilliQ water and concentrations of $1\,mmol\,l^{-1}$, $2\,mmol\,l^{-1}$, $3\,mmol\,l^{-1}$, $4\,mmol\,l^{-1}$

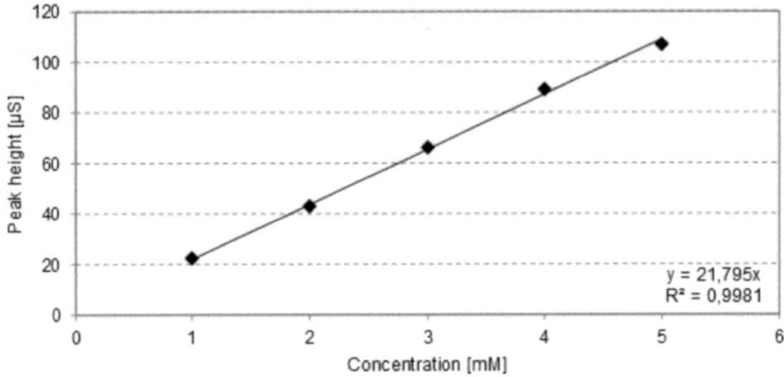

Fig. 3.12: Calibration curve for sulfate ion analyzed by ion chromatograph Dionex.

and $5\,mmol\,l^{-1}$ were analyzed. The principle of the test was to measure the turbidity change of a sample, as the sulfate was precipitating as barium salt $BaSO_4$ after $BaCl_2$ solution addition. Calibration curves of an acceptable regression coefficient were obtained (R^2 = 0.959). However, in comparison to ionic chromatography, it was less accurate.

3.6 Protein concentration measurement

Protein concentration was measured spectrophotometrically with a Hach Lange DR 5000 spectrophotometer according to Lowry et al. (1951). Calibration with serum bovine albumin suspended in water and in 1 M NaOH as a standard solution gave a very good regression coefficient for different concentrations in both cases (Figure 3.13). NaOH solution was applied as a solvent because of the protein extraction method for sample preparation. Further, 1 ml sample was collected for this analysis and centrifuged at 15000 rpm for 5 min . The liquid phase was decanted and the remaining biomass pellet was suspended in 5 ml 1 M NaOH for protein extraction at 95 °C for 10 min in an Eppendorf Thermomixer 5436 with slight stirring. The time of incubation at elevated temperature was determined experimentally (Figure 3.14).

Protein concentration measurement was applied for biomass growth control in a consortium, where microbial aggregates formation was observed. The above procedure resulted from the following test: Disintegration with an ultrasonic bath was performed, to test if mechanical disintegration of flocks/ aggregates was possible. Two samples of pre-grown cultures with the aggregates formed were analyzed. Incubation in the ultrasonic bath was attained in 15 min and

30 min time. The optical density was measured and the results are presented in the Figure 3.15. The difference in the measured extinction was slightly noticeable, so the disintegration was not successful, and the protein determination according to Lowry et al. (1951) was applied for bacterial growth monitoring in ecosystems with aggregate formation instead of direct spectrophotometric analysis (OD).

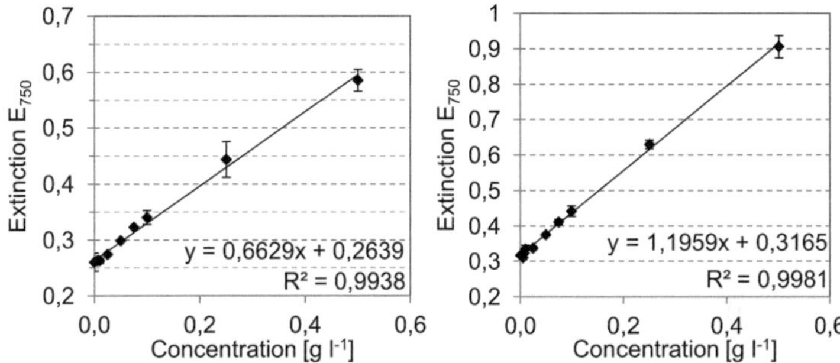

Fig. 3.13: Calibration curves prepared for photometric protein analysis at $\lambda = 750$ nm with serum albumin bovine as a model protein, with deionised water (left plot) and 1 M NaOH as a solvent (right plot). Performed according to Lowry (1951).

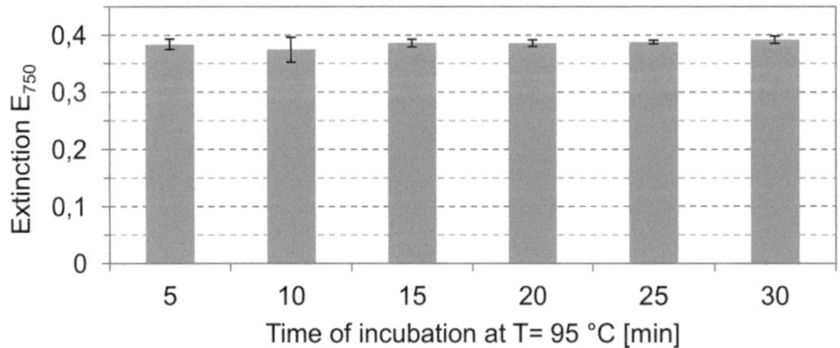

Fig. 3.14: Experimentally determined incubation time for protein extraction with 1 M NaOH solution at elevated temperature.

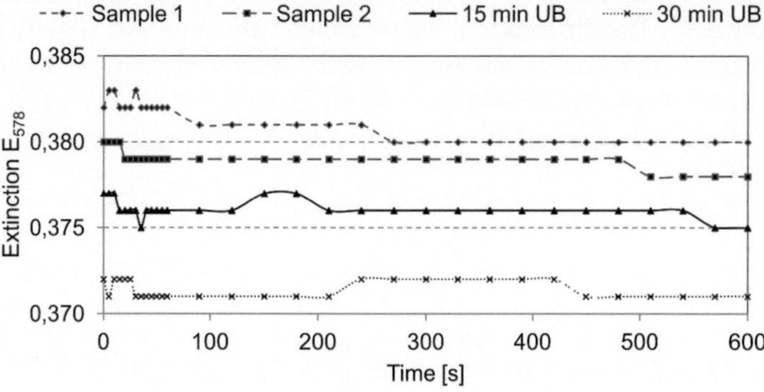

Fig. 3.15: Experimentally determined incubation time for protein extraction with 1 M NaOH solution at elevated temperature.

3.7 Optical density as a growth control

Optical density at a $\lambda = 578\,$nm was measured with a photometer for biomass growth monitoring.

3.8 Formate measurement with an enzymatic test

Formic acid was measured with the use of a formate dehydrogenase test kit of Boehringer (Mannheim, Germany). The detection limit was $0.02\,$mmol$\,$l^{-1}. Measurement principle was the NADH amount increase by light absorbance change at $\lambda = 340\,$nm, according to (3.1) (Höpner et al. 1974). The increase of NADH concentration is proportional to the concentration of formate in the sample. A standard solution was prepared with fumaric acid disodium salt (p.a.) purchased by Merck-Schuchardt. The concentration was $0.1\,$g$\,$l^{-1}.

$$HCOO^- + NAD^+ + H_2O \xrightarrow{FDH} HCO_3^- + H^+ + NADH \qquad (3.1)$$

3.9 Statistical analysis

Statistical analysis- one or two-way analysis of variance (ANOVA) and curve fittings were performed with the statistic program Graph Pad Prism 5.0 (GraphPad Software, Inc., USA). All results are given with an error calculated as a standard deviation for n > 2 measurements.

3.10 Microscopy

During all experiments microscopic observations were one of the most important control approaches. Changes in microbial population during the growth in samples of different origin and in microbial enrichments on propionate were conducted with a microscope. Some other tests like, e.g. spore staining or fluorescence *in-situ* hybridization were also evaluated with the use of microscopy.

3.10.1 Used microscopes

An epifluorescent microscope Zeiss Axioskop A50 equipped with a mercury HBO 50 UV lamp was used for routine analysis. Filter sets listed in Table 3.5 were applied during FISH experiments and for routine control of the activity of methane bacteria.

Table 3.5: Filter sets implemented for the microscopic observation under UV light.

Filter set	05	09	TRITC	DAPI
Excitation	BP 395-440	BP 450-490	Brightline HC 543/22	Brightline HC 377/50
Beam Splitter	FT 460	FT 510	HC BS 562	HC BS 409
Emission	LP 470	LP 515	Brightline HC 593/40	Brightline HC 447/60
Application	F_{420} observation	FAM label observation	Cy3 label observation	DAPI staining observation
Brand	Carl Zeiss Microscopy GmbH	Carl Zeiss Microscopy GmbH	AHF Analysentechnik Tübingen	AHF Analysentechnik Tübingen

3D pictures of microbial aggregates were obtained with a Carl Zeiss microscope equipped with an ApoTom function (Axio Imager Z1, Zeiss, Jena, Germany). The following filter sets were embedded: DAPI, FITC (FAM), Cy3 and Cy5 for observation of fluorescent labels of the same name.

All microscopic images presented in this work were taken after application of immersion oil onto the covering glass of the specimen and using a magnification factor of 1000.

3.10.2 Applied fluorescent dyes and oligonucleotide probes

For identification purposes 16S rRNA-based oligonucleotide probes were applied. All of them are listed, together with representative dyes and the target organism group, in Table 3.6. The probes were purchased as modified oligos á la Carte at Eurofins MWG Operon (Munich, Germany).

Table 3.6: Specific oligonucleotide probes used during the research.

Probe	Sequence (5'-3')	Specifity	Label	Ref.
Eub 338	GCT GCC TCC CGT AGG AGT	most *Bacteria*	5'-Cy3 5'-FAM	Amann et al. 1990
Arc 915	GTG CTC CCC CGC CAA TTC CT	*Archaea*	5'-Cy3 5'-FAM	Stahl and Amann 1991
MPOB1	ACG CAG GCC CAT CCC CGA A	*Syntrophobacter fumaroxidans* and *Syntrophobacter pfennigii*	5'-FAM	Harmsen et al. 1998
KOP1	TCA AGT CCC CAG TCT CTT CGA	*Syntrophobacter pfennigii*	5'-Cy3	Harmsen et al. 1998
Synbac 824	GTA CCC GCT ACA CCT AGT	*Syntrophobacter*	5'-Cy3	Ariesyady et al. 2007
Smi SR 354	CGC AAT ATT CCT CAC TGC	*Syntrophus* group incl. *Smithella propionica*	5'-FAM	Ariesyady et al. 2007
SRB 385	CGG CGT CGC TGC GTC AGG	most *Desulfovibrionales* and other sulfate reducing bacteria	5'-FAM	Amann et al. 1990
Syb 701	AAA TGC AGT TTC CAA TGC AC	*Syntrophobacter* related (MUG28 clone)	5'-Cy3	Sekiguchi et al. 1999
Delta 495	AGT TAG CCG GTG CTT CCT	*Deltaproteobacteria*	5'-DY415	Loy et al. 2002, Lücker et al. 2007

3.10.3 Spore staining

Spore staining was applied to characterize bacteria in terms of their ability of spore formation. Malachite green and safranin were used for the dyeing procedure. The solutions were:

A) Malachite green solution:

- 5 g malachite green oxalate (p.a.)
- 100 ml distilled water

B) Safranin solution:

- Stock solution:

  3.41 g safranin-O (p.a.)

  100 ml 96 % ethanol (p.a.)

- Working solution:

  10 ml stock solution

  100 ml distilled water

The procedure started with spreading of a bacteria suspension onto microscopic slides followed by air drying and heat fixation. The specimen was afterwards coverd with malachite green solution and placed over a water bath (steaming water in a small beaker) and incubated in water vapor for 5 minutes, reapplying the dye if it started to dry out. The slide was than washed with distilled water until the washings run clear and the organisms were counterstained with safranin for 20 seconds. After rinsing the slide was dried and observed under oil immersion with a light microscope. Control samples were prepared with *Smithella propionica* (Figure 3.16).

3.10.4 Gram staining

For determination whether organisms were Gram-positive or Gram-negative, or in other words if their cell wall contains a thick layer of murein (or pseudomurein), two dyes were applied, then Lugol's solution and 96 % ethanol. Solutions used:

A) Crystal violet- ammonium oxalate solution:

- Solution a:

  13.87 g crystal violet (p.a.)

  100 ml distilled water

- Solution b:

  0.8 g ammonium oxalate monohydrate (p.a.)

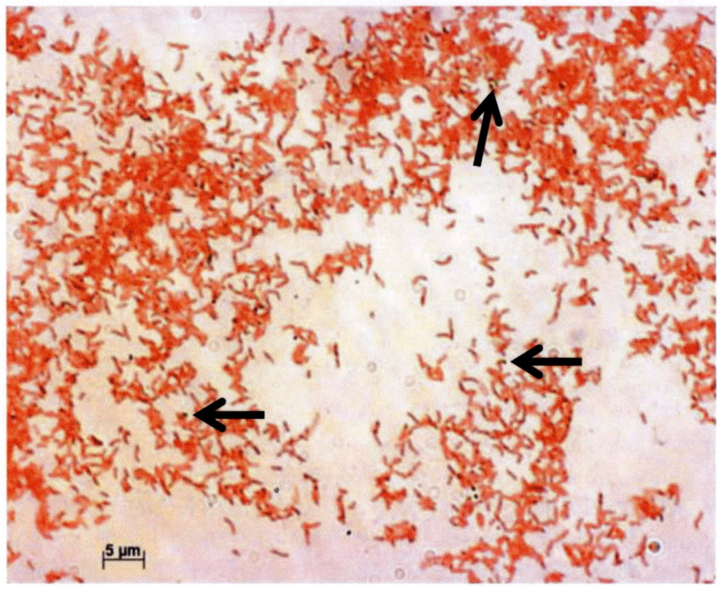

Fig. 3.16: Control spore staining of *Smithella propionica* arrows point to green spores.

   80 ml distilled water

- Dye solution:

   20 ml solution a

   80 ml solution b

B) Lugol's solution:

- 1 g iodine (p.a.)

- 2 g potassium iodide (p.a.)

- 300 ml distilled water

Safranin was prepared as described in the section 3.10.3.

The staining procedure was as follows: A sample containing bacteria was first spread onto a microscopic slide, air-dryed and heat fixed. Crystal violet ammoniumoxalate solution was put onto specimen and after 2 min washed with Lugol's solution. Specimen was covered with Lugol's solution for 2 min and washed afterwards with 96 % ethanol until washings were colorless. Safranin solution was then applied and incubated for 20 s on a specimen. Afterwards the specimen was washed with distilled water, dried and taken for microscopic observation. Control staining was prepared with *Enterococcus* (Gram-positive, dark blue) and Smithella propionica (Gram- negative, red) (Figure 3.17).

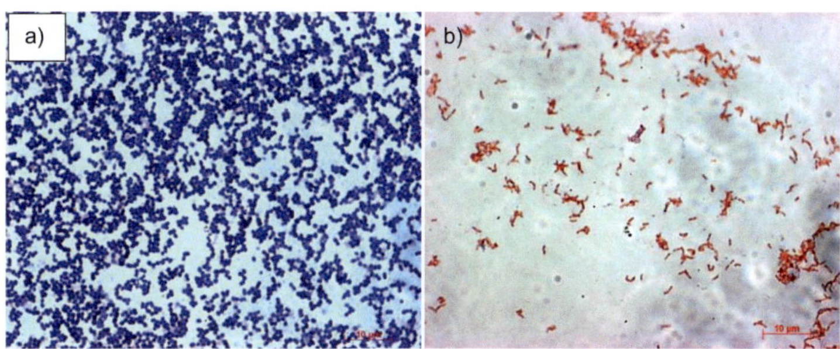

Fig. 3.17: Microscopic representation of control Gram staining for a) Gram-positive *Enterococcus* (dark blue) and b) Gram-negative *Smithella propionica* (red).

3.10.5 Fluorescence *in-situ* hybridization

The 16S rRNA-based oligonucleotide probes listed in Table 3.6 were used for FISH analysis. Fixation of microorganisms was performed according to Amann et al. (1990). Oligonucleotide probes labeled with a fluorescent dye at 5'-end were used for hybridization.

Hybridization buffer, prepared always fresh in 2 ml Eppendorf reaction tube, consisted of 360 µm 5 M NaCl, 40 µm 1 M Tris/ HCl pH 8.0; 30 % formamide (600 µm), 1 ml ultra pure water and 4 µm of 10 % (w/ v) SDS. This buffer was dropped at each sample well on the teflon coated glass slide (8 µl to 10 µl). Afterwards appropriately diluted probes were applied (4 µl). The specimen was then transferred into a hybridization tube (50 ml Falcon tube) and incubated for 90 min in an oven at a temperature of 46 °C.

The washing step came after hybridization. It was performed with the use of washing buffer prepared in a 50 ml Falcone tube, preheated in a water bath at 48 °C before use. It contained 1 ml 1 M Tris/ HCl pH 8.0; 1 ml 5 M NaCl, 500 µm 0.5 M EDTA and 50 µm 10 % (w/ v) SDS. The slide was firstly washed with a small amount of warm buffer, and next incubated in a Falcon tube (containing buffer remains) for 20 min.

Before the microscopic observation, the specimen was counterstained with $0.1\,\mu\mathrm{mol\,l}^{-1}$ 4',6-diamidino-2-phenylindol (DAPI). DAPI is an universal dye, which binds strongly to DNA and gives a blue color to all organisms in the sample. The process was performed in darkness, on ice for 15 min. After drying the slide, samples were covered with embedding agent CitiFluorTM (glycerol/ PBS solution AF1, London, United Kingdom) and observed by means of fluorescent microscopy.

A) 5M NaCl

- NaCl (29.22 g)
- H_2O (100 ml)

B) 1M Tris/ HCl

- Tris (tris-(hydroxymethyl)-aminomethane) 12.11 g
- H_2O 100 ml

  pH was adjusted with HCl to 7.2- 7.4.

C) 10 % SDS

- SDS (sodium dodecyl sulfate) (10 g)
- H_2O (100 ml)

  First the glass was filled with ultra pure water up to 80 ml, and as soon as SDS dissolved, the solution was filled up to 100 ml. Sterile filtration with the filter of 0.2 µm was performed.

D) 0.5 M EDTA pH 8.0

- EDTA (ethylenediaminetetraacetic acid) (14.61 g)
- H_2O (100 ml))

  pH was adjusted to 8.0.

E) Sodium phosphate buffer

a) Solution A:

- Na_2HPO_4 (35.60 g)
- H_2O (1000 ml)

b) Solution B:

- NaH_2PO_4 (6.9 g)
- H_2O (250 ml)

pH of solution A was adjusted with solution B to pH 7.2- 7.4.

F) 1xPBS

- NaCl (7.6 g)

- Sodium phosphate buffer (50 ml)

- H_2O (filled up to 1000 ml)

G) 3xPBS

- NaCl (22.8 g)

- Sodium phosphate buffer (150 ml)

- H_2O (filled up to 1000 ml)

H) PFA

- PFA (paraformaldehyde) (4 g)

- 2M NaOH (few drops)

- 3xPBS (33 ml)

  65 ml MilliQ water was heated to 60 °C, and then PFA was added together with one drop of 2M NaOH and stirred rapidly until the solution clarified (1 min to 2 min). If it did not clarify, more drops of NaOH were added. The solution was removed from the heater, 33 ml 3xPBS were added and the pH was checked. Sterile filtration was done with 0.2 μm filter. After the solution cooled down; it was distributed into 2 ml Eppendorf reaction tubes with 1.5 ml PFA, frozen and stored at -20 °C.

3.10.6 Specimen observation and microscopic images' analysis

Three dimensional images taken with ApoTome
for interspecies distances measurements
Microscopic pictures taken in order to get a 3D aggregate image, with a z-stock dimension set on 0.3 μm. The AxioVision software (Carl Zeiss Microimaging GmbH, Jena, Germany) was applied for microscopic picture analyses. The distances between differently stained organisms were measured with image analysis software with all z-stack images taken into consideration. Measurements were done with an accuracy of 0.01 μm. At each of the growth stages the span was measured from the cell identified as a propionate

degrading organism to the Arc 915-hybridized cells positioned in the closest vicinity, giving from 3 up to 15 measurements per single image. That makes up to ca. 1000 lengths averaged to give one value of a distance at each growth phase. Additionally the aggregates dimensions were measured: brightness (x), height (y) and thickness (z) (Figure 3.18). These values were also averaged upon each image, containing usually one to two aggregates, taken during each measurement session.

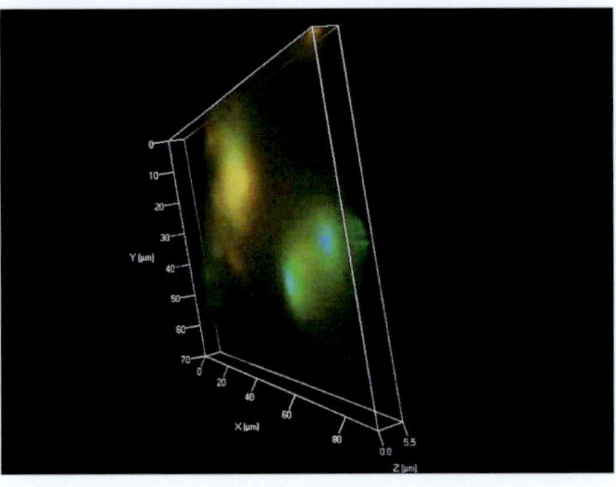

Fig. 3.18: 3D aggregate image photographed with an ApoTome (Axio Imager Z1, Zeiss, Jena, Germany) in an AxioVision software with dimensions defined as x for brightness, y for height and z for thickness.

Population analysis

Population change during the growth of microorganisms was evaluated with the daime software for digital image analysis in microbial ecology (Daims et al. 2006). Microscopic pictures were used for biovolume fraction evaluation, relying on the comparison of a total biomass biovolume stained with unspecific fluorescent dye (DAPI) and the biovolume fraction of organisms population labeled with 16S rRNA targeted oligonucleotide probe. Although DAPI is not recommended for such kind of evaluation (Nielsen et al. 2009), using it as a base for population fluctuations description at each growth phase gives a reliable overview of population ratio changes within examined ecosystem.

4 Results

4.1 HS-GC-MS method for volatile fatty acids determination – development and application

Operating parameters of the combined headspace - gas chromatography - mass spectrometry system (HS-GC-MS, Figure 3.5) were experimentally varied to obtain the highest values of analytes' peak areas for specified concentrations, giving a calibration curve of acceptable regression coefficients and appropriate fragmentation ions in the mass spectra. Most suitable parameters were developed through operational parameters for equipment and sample preparation approach optimization.

4.1.1 Sample preparation

After about 30 days growth of biomass enriched from several digesters (section 4.4), when appropriate concentration of microorganisms was reached, the sample was transferred into anaerobic chamber (COY Laboratory Products Inc., Ann Arbor, USA) with an atmosphere of N_2 (95 %) and H_2 (5 %). 2 ml portions were centrifuged in an Eppendorf Minifuge at 13000 rpm for 2 min. The liquid supernatant was decanted and the bacterial pellets were washed 3 times with a freshly prepared medium (pH 7) and finally suspended in the 10 ml broth, containing ^{13}C-propionate. This approach was chosen to avoid the interference of residual VFAs with ^{13}C-labeled propionic acid degradation and to accelerate the degradation of propionic acid by concentrated biomass. The concentrated mixed bacteria were transferred into a serum bottle (usually 60 ml in total volume) and supplemented with ^{13}C labeled sodium propionate. The bottles were immediately closed with a rubber stopper which was secured with an aluminum cap and incubated at 37 °C in darkness. During the incubation time 100 µl portions of the suspension were regularly withdrawn with syringes until propionate was completely degraded. The samples were diluted with $NaHSO_4 \cdot H_2O$ (50 % w/v, 1:1 ratio) and centrifuged in Eppendorf cups (1.5 ml) at 13000 rpm for 2 min before subjected to the HS-GC-MS system.

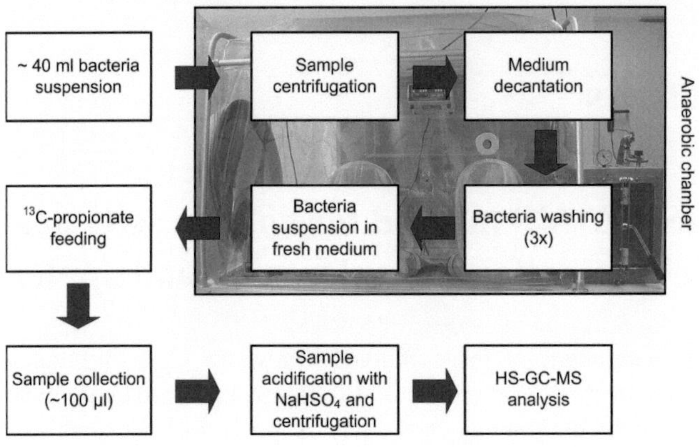

Fig. 4.1: Working procedure for sample preparation before HS-GC-MS analysis.

Forty µl supernatant was transferred into a glass vial (22 ml), closed with a teflon-coated silicone rubber septum and secured with an aluminum cap. The whole working procedure for the ^{13}C-propionate experiments is summarized in Figure 4.1.

Before elucidation of this approach, samples collected during ^{13}C-labeled propionic acid degradation were first centrifuged and then acidified. This resulted in obtaining smaller peaks areas for analytes. Comparison of peak areas for differently prepared samples is shown in Figure 4.2. Acidification of collected liquid with organisms present gave more accurate values due to lowering pH of the suspension, and hence, forcing the acids to diffuse through the cell wall. This allowed more accurate determination of the actual mass balance between substrates and products during microbial conversions.

According to the basic procedure of volatile fatty aacids determination applied since 1981 (Hoshika), the acidifying agent applied most often for gas chromatographic separation was hydrophosphoric acid. This was the optimal solution, as the columns packed into Teflon coating usually were equipped with Carbopack C modified with FFAP and H_3PO_4. The H_3PO_4 was also applied during the first steps of method optimization for the purpose set in the goal of this thesis, however better analytes' recovery was obtained with $NaHSO_4$ solution, what will be discussed in the section 4.1.2.

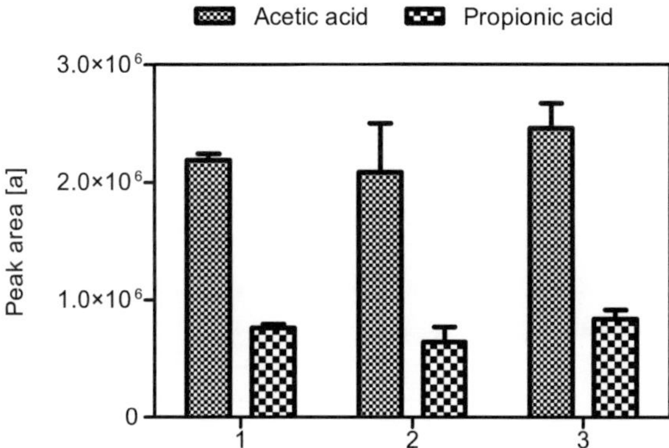

Fig. 4.2: Resulting peak areas of acetic and propionic acid for the same sample of mixed culture during growth on propionic acid for different preparation procedures: 1- sample centrifugation, biomass pellet rejection and supernatant acidification; 2- acidified sample with biomass; 3- sample acidification, centrifugation and supernatant collection.

4.1.2 Parameters optimization

Headspace Sampling unit Dani 3950

The headspace sampling unit Dani 3950 was equipped with a silicone oil bath with temperature adjustable in the range of 40 °C to 150 °C, a sample holding carousel with 24 positions, a 1 ml sampling loop and a transfer line maintained at the temperature of 120 °C (not adjustable). Sampling occurred with a syringe penetrating the sample vial secured with a teflon coated septum, pressurizing (10 s) the vial with He carrier gas and then collecting 1 ml of the vapors-He gas mixture flowing through the HS at a 1 ml/ min flow rate.

The time of sample incubation in the silicone oil bath, prior to sample collection, the bath temperature and sample volume were experimentally adjusted to obtain the highest peak area.

The range of temperature for silicon-oil bath in the HS was 40 °C to 150 °C. For the purposes of this study the value of 80 °C was set as optimal since the examined samples contained a large content of water (setting the temperature at an appropriate level prevented from introducing elevated concentration of water vapor into the GC-MS system). This was so to get a stable environment for ionization in the ionic chamber of MS and hence, good peaks resolution. Secondly, the boiling points for acetic acid and propionic acid are, respectively, 118 °C and 141 °C, so the temperature chosen had to be sufficiently high in order to bring possibly the largest fraction of an analyte into headspace. Since both acids are volatile, this criterion was fulfilled at a temperature of 80 °C.

The significance of the sample volume was also examined experimentally, starting with 5 ml, 3 ml and 2 ml liquid in 22 ml glass vial. This approach resulted in very poor resolution of produced chromatograms. The sample was first incubated at 80 °C to obtain the VFAs in the gaseous form. In such case the time of incubation should be sufficiently long, to get satisfactory peak area, for the volumes of 2 and 3 ml, the time set was 30 min. Nevertheless, the longer the incubation time, the larger the water vapor content in the head-space, what resulted in elevated baseline read out (up to 100 kCounts by 3 ml volume) and very poor chromatogram resolution. The peaks in the mass spectra were not corresponding with the ones from the NIST library, giving no matches during the library search. Out of several studies, further approaches were tested. To meet the goal a total vaporization technique (TVT) was used (Kolb and Ettre 1997). The technique is based on applying small volumes of liquid samples into large glass vials. The volume tested was 20 µl in the 22 ml vial. The acidifying agent used was 4 % (v/ v) hydrophosphoric acid. The incubation time was 15 min and the temperature was set at 80 °C. The examined VFAs mixture was composed of acetic acid, propionic acid, butyric acid, iso-butyric acid, valeric acid and iso-valeric acid at a concentration of 5 mmol l^{-1}, each analysis resulted in a chromatogram with six different peaks of good separated acids and an additional peak of phthalate with a 149 m/ z as a 100 % intensity ion in the mass spectrum (Figure 4.3). The contamination was caused probably by the butyl rubber septa used to seal the glass vials. Changing the septa to teflon-coated silicon rubber septa diminished the phthalate pollution in the system.

At this step of method elucidation, the chromatogram of a good resolution was obtained, whereby the peaks were successfully separated and detected. However, the ions for several VFAs were not corresponding with the ones in the library. As stated previously, lowering the pH of the analyzed sample and salt addition enhanced the extraction of acids from liquid into gaseous phase (Pan et al. 1995). This was the reason for the search of an optimal matrix for the sample preparation. The NaCl addition decreases solubility of acids in the nondissociated form, increasing the Henry constant and, hence, the gas phase transfer. The agent that meets both, the requirement for lowering pH and salt addition effect is $NaHSO_4$. The hydrated form was applied for further analysis, and a saturated solution (50 % w/ v) was prepared in MilliQ water. The pKa of $NaHSO_4$ is relatively low (1.9) and lowers pH of analyzed

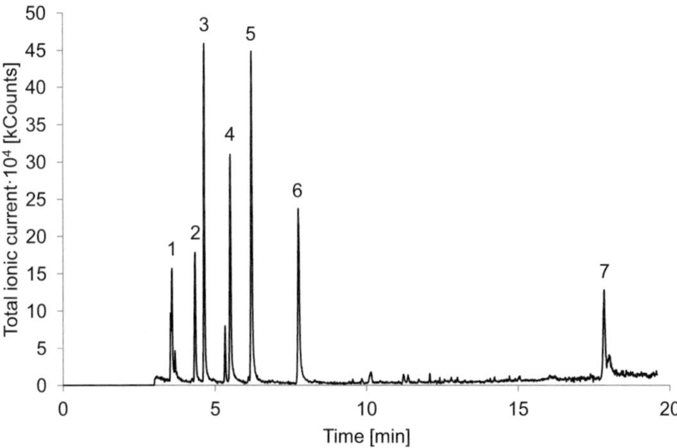

Fig. 4.3: Chromatogram when Total Vaporization Technique was applied, MS identified peaks: 1) acetic acid, 2) propionic acid, 3) iso-butyric acid, 4) butyric acid, 5) iso-valeric acid, 6) valeric acid and 7) phthalate contamination. Concentration tested $5\,mmol\,l^{-1}$; acidifying agent used H_3PO_4; split ratio 15.

samples down to 1 if mixed at a 1:1 ratio. The finally selected volumes for the analysis were 20 µl sample and 20 µl $NaHSO_4$ solution. The effect of the acidifying agent on acetic acid and propionic acid peak areas from standard solutions of the same concentration ($5\,mmol\,l^{-1}$) are represented in Figure 4.4.

Analytes were than transferred further into the gas chromatograph, where the temperature program was chosen.

Gas Chromatograph Varian 431C

The GC was connected with the HS Dani with a heated transfer line ending with a steel needle that was directly introduced into the GC injector. The transfer line temperature was set to 120 °C and there was no possibility of its adjustment. Adequate injector temperature (here 250 °C), optimal flow rate of inert carrier gas and an optimal temperature program, for GC were the most important in obtaining chromatograms of good resolution, whith narrow, non-tailing peaks. Carrier gas was 5.0 He (99.999 % purity); the flow rate was set to 1.0 ml/min.

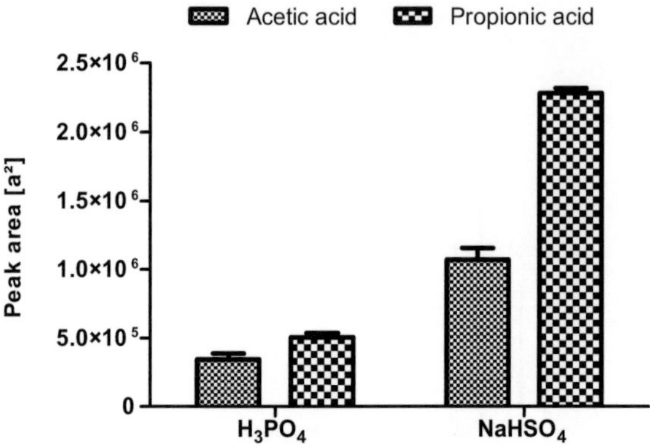

Fig. 4.4: Differences between peaks areas for $5\,\mathrm{mmol\,l^{-1}}$ acetic- and propionic acid caused by sample matrix change.

The GC was equipped with an 1177 split/ splitless injector with electronic flow control, which monitors and adjusts the pressure of the carrier gas introduced to the system. The back pressure regulator controls the pressure at the head of the column. The split ratio represents the fraction of the sample that enters the column and is adjustable adequately to the analysis needs. Setting it to 10 means that $(\frac{1}{10})^{th}$ of the sample enters the column. As the sampling loop of HS Dani had a volume of 1 ml, the split function had to be applied. Otherwise column overloading was observed. Both options 15 and 10 were examined with finally acquired value of 10, giving $100\,\mu l$ of gaseous sample entering the system. While setting the splitless mode the water content was so high, that no peak separation was observed and the basieline was recorded at the level exceeding 200 kCounts, what was more than the read out for the propionic acid highest peak point when the split ratio was set to 15 (Figure 4.5). For the curve obtained with splitless mode, the 100 % relative intensity peak was placed at the position of 18 m/ z, what corresponds with water ion (H_2O^+). Within the time the water vapor content was increasing in the system, so to maintain good analysis conditions final adjustment was the split ratio of 10.

A temperature program for the analysis of VFAs was chosen to obtain the best separation of analytes. The starting point was $80\,°C$, as this was the sample incubation temperature in the silicon oil bath of HS, going through the different degrees rate increase in time, ending at $150\,°C$, for acetic acid and propionic acid determination and at $200\,°C$ for VFAs with a carbon chain length up to

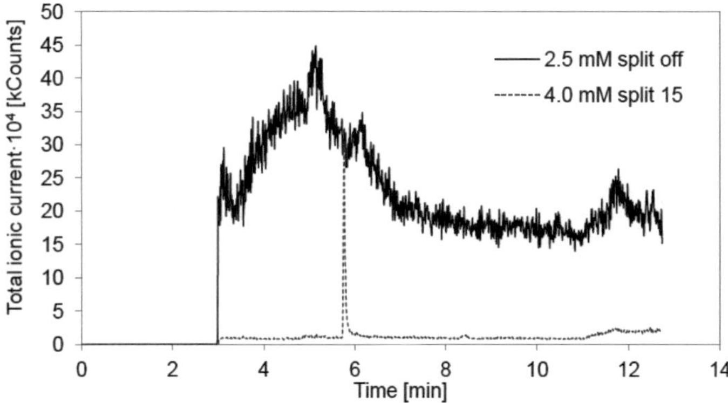

Fig. 4.5: Examples of chromatograms obtained for split and splitless mode with a standard containing propionic acid at 2.5 mmol l^{-1} and 4.0 mmol l^{-1} concentration. mM = mmol l^{-1}

five C atoms (Table 9). The appropriate program provides the shortest time of chromatographic separation of analytes, allowing qualitative and quantitative analysis of the compounds present in the sample. The separation results are presented in Figure 27 (five VFAs) and Figure 30 (acetic and propionic acid).

Mass Spectrometer Varian 210

The parameters for the MS were experimentally acquired to obtain the m/ z values corresponding with exact molar masses of analytes in the mass spectra. Such spectra could be interpreted immediately, securing rapid data processing.

The parameters that were acquired to obtain exact distribution of mass intensities separation were:

- Axial modulation (AM)- acquired according to the assumption that the lowest possible value providing good resolution of 131/ 132 m/ z and 414/ 415 m/ z calibration gas peaks is desired, the range of inspection was 2.0 V to 4.0 V; in case when the AM is too low, high molecular ions will be missing, and if AM is too high the peak width for low molecular ions will be broadened and mass miss assignments might occur; during method optimization lowering the AM value was checked experimentally for the best resolution, starting from 4.0 V; the optimal value for set requirements was established to be 2.2 V; setting the value

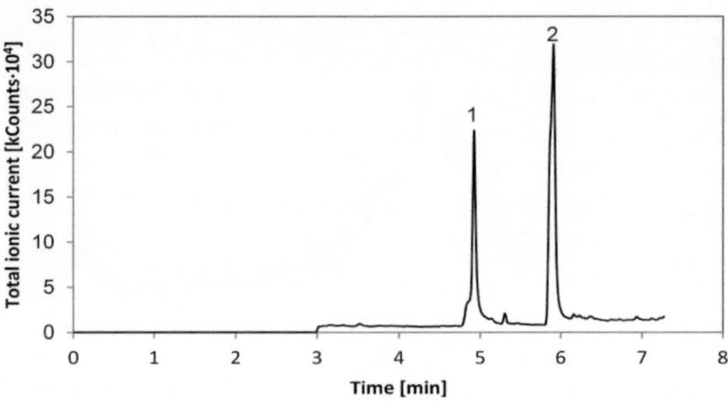

Fig. 4.6: Acetic- and propionic acid (C2, C3) after chromatographic separation with a temperature program 80 °C(1 min) $\longrightarrow$ 8 °C/ min $\longrightarrow$ 120 °C $\longrightarrow$ 150 °C(1 min); concentration 5 mmol l^{-1}; the acidifying agent used was NaHSO$_4$; split ratio 10.

below 2.2 did not result in calibration failure of the ionic chamber, howeverthe MS auto-tuning was unsuccessful.

- Emission current- Filament emission current increases the number of ions produced. The chosen value was 10 µA; although changing to higher values had no influence on the main fragmentation ions shift for acetic acid, it resulted in production of molecular ions for propionic acid at a position of 72.8 m/ z (20 µA and 73.0 m/ z (40 µA), (compare Figure 15 a).

- Ion trap temperature- Setting the temperature at the range between 150 °C and 175 °C extends the lifetime of the turbomolecular pump. For straight chain carbohydrates it is recommended to apply hotter conditions, which was not the case as relatively large content of water vapor was observed in used system (due to liquid sample incubation at 80 °C, the temperature of 180 °C was set.

Combinations of some parameters, with the influence on the fragmentation ions for acetic/ and propionic acid are listed in Table 4.1. MS ionization was set to EI (70 eV), the scanning was performed in the SCAN mode, analyzing the whole mass spectrum (10 m/ z- 650 m/ z). The exact fragmentation ions, matching the mass spectra from NIST library (Figure 15 a, b), were found for the setting 10 µA ionization current and 2.2 V axial modulation.

For further analysis the SCAN mode was switched to SIM mode (selected ion monitoring) to obtain better selectivity of the method. The range was set to

Table 4.1: Settings for the MS and their influence on ionization and fragmentation ions production.

Parameter	Setting					
Emission current [μA]	10	10	20	10	40	10
Axial modulation [V]	4.0	3.5	3.5	3.0	3.0	2.2
Ionisation time [μs]	5000	25000	25000	25000	25000	25000
PRI/ FI[1]	I[2] II[3]	I[2] II[3]	I[2] II[3]	I[2] II[3]	I[2] II[3]	I[2] II[3]
Acetic acid	44.9 59.8	45.0 59.8	45.0 59.8	42.9 61.0	43.0 59.8	43.0 59.9
Propionic acid	44.9 73.9	45.0 72.8	45.0 72.8	57.0 74.9	44.9 73.0	45.0[4] 74.0

[1] PRI = peak relative intensity; FI = fragmentation ion
[2] 100 % relative intensity fragmentation ion [m/ z]
[3] Molecular ion [m/ z]
[4] 95.1 % relative intensity fragmentation ion [m/ z]

35 m/ z- 100 m/ z. That eliminated the problem of high water vapor content and phthalate contamination influence on peaks detection. The obtained mass spectra of acetic acid and propionic acid are presented in Figure 4.7.

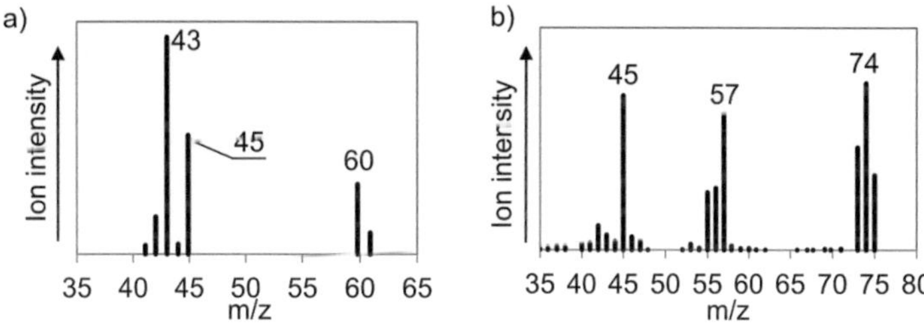

Fig. 4.7: Mass spectra obtained after parameters optimization for a) acetic acid and b) propionic acid.

4.2 Detection and differentiation of ^{13}C-labeled metabolites

For propionic acid degradation pathway identification, the check on the behavior of ^{13}C- labeled acetate (product of ^{13}C-labeled propionate degradation) and ^{13}C-labeled propionate had to be analyzed for the parameters acquired during its optimization. Calibration curves for 1-^{13}C-labeled acetic acid and for 2-^{13}C-labeled homologue were prepared. The first one is ^{13}C labeled at the carboxylic moiety, and the second at the methyl moiety of a compound. Acids used were both sodium salts purchased from Sigma Aldrich. Standard solutions were prepared with MilliQ ultrapure water and acidified with 50 % $NaHSO_4 \cdot H_2O$ at a ratio 1:1, according to the sample preparation procedure described before.

Both the calibration curves and chromatograms obtained experimentally were satisfactory, giving good regression coefficients and the expected fragmentation ions (Table 3.2) in the mass spectra. Calibration curves are represented in Figure 4.8. The mass spectra obtained for 1-^{13}C-acetic acid and 2-^{13}C-acetic acid are presented in Figure 4.9. The fragmentation ions corresponded to the predicted ones (Table 3.2).

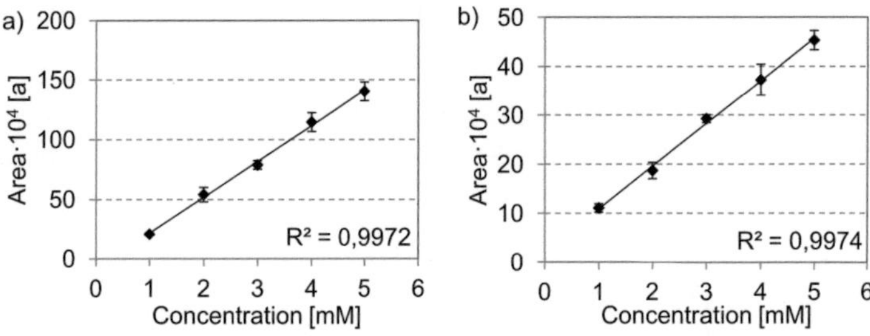

Fig. 4.8: Calibration curves for a) 1-^{13}C-labeled acetic acid and b) 2-^{13}C-labeled acetic acid; mM = mmol l^{-1}.

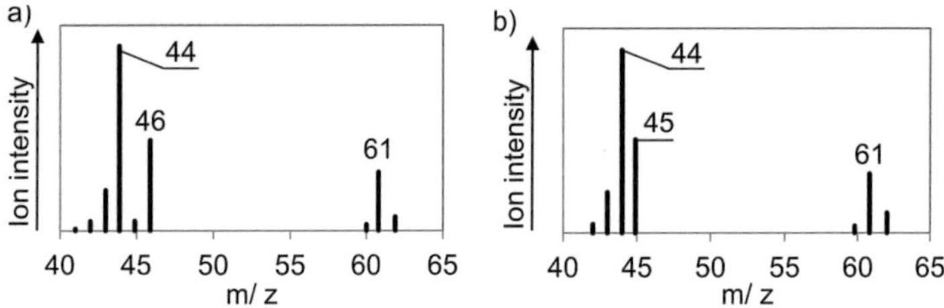

Fig. 4.9: Mass spectra obtained for a) 1-^{13}C-labeled acetic acid and b) 2-^{13}C-labeled acetic acid.

The substrate used for pathway analysis was 1-^{13}C-labeled propionic acid sodium salt, for which the fragmentation ions (Figure 4.10) were congruent with ones listed in Table 3.2. A specification of all used chemical compounds is listed in Table 4.2.

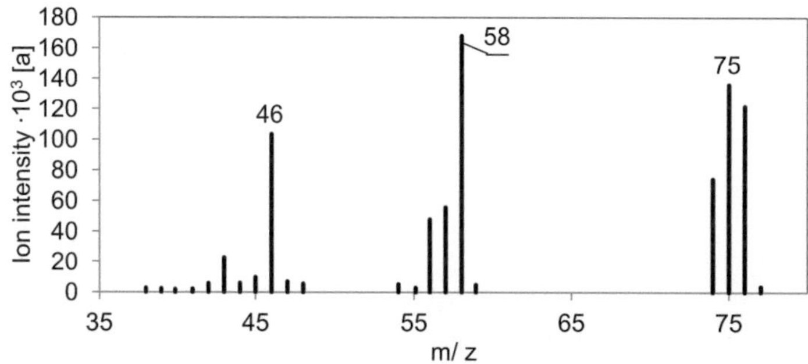

Fig. 4.10: Mass spectrum of 1-^{13}C-labeled propionic acid.

Table 4.2: ^{13}C-labeled substances used during the study.

Chemical name	CAS number	Chemical formula	^{13}C [%]	Manufacturer	Purity
Sodium acetate-1-^{13}C	23424-28-4	$CH_3^{13}COONa$	99	IsotecTM, Ohio, USA	p.a.
Sodium acetate-2-^{13}C	13291-89-9	$^{13}CH_3COONa$	99	IsotecTM, Ohio, USA	p.a.
Sodium pro-pionate-1-^{13}C	62601-06-3	$CH_3CH_2^{13}COONa$	99	Sigma-Aldrich, Taufkirchen, Germany	p.a.

4.2.1 Final parameters and method summary

The overall method parameters for each of the system components have been elucidated and summarized in Table 4.3. Calibration curves were prepared for metabolites planned to be analyzed. Standard solutions prepared from acetic acid anhydride (99.5 %, p.a.) and propionic acid (99 %, p.a.) both from Fluka, Taufkirchen (Germany) at concentrations of 1 mmol l^{-1}, 2 mmol l^{-1}, 3 mmol l^{-1}, 4 mmol l^{-1} and 5 mmol l^{-1} were analyzed. The resulting regression for both analytes is represented in Figure 35.

Determination of limits of detection (LOD) was the next step of method development. These values were determined for acetic and propionic acid using 20 blank samples (according to (4.1)). The method was chosen because the traces of both acids were measurable without introducing the standard into prepared samples. The obtained LOD values were 0.09 mmol l^{-1} for acietic acid and 0.02 mmol l^{-1} for propionic acid.

Table 4.3: Experimentally optimized parameters for the HS-GC-MS system for metabolic pathway analysis.

HS - DANI 3950	GC - Varian 431C	MS - Varian 210
Liquid volume: 40 µl	Injector temperature: 250 °C	Ionization mode: EI
Bath temperature: 80 °C	Split ratio: 1:10	Ionization energy: 70 eV
Incubation time: 15 min	Temperature program:	Trap temperature: 180 °C
Gas sampling time: 10 s	$80\,°C \xrightarrow{8\,°C/min} 120\,°C \longrightarrow$	Axial modulation: 2.2 V
Gas sample volume: 1 ml	$\xrightarrow{40\,°C/min} 150\,°C$	
	Carrier gas: He (5.0)	
	Carrier gas flow: 1.0 ml/min	
	Total analysis time: 7 min 30 s	

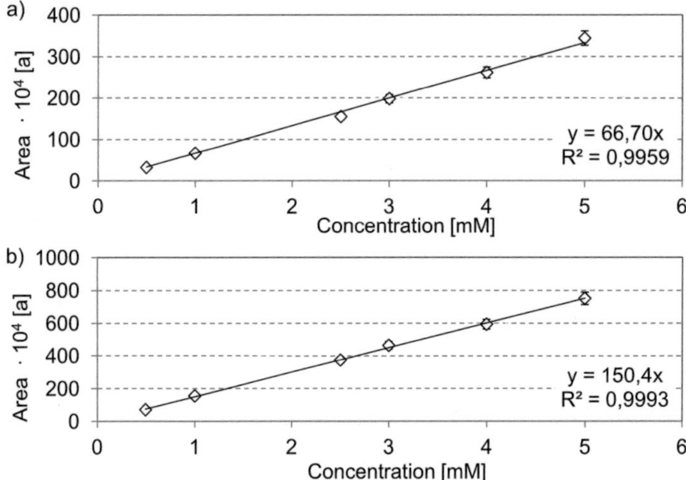

Fig. 4.11: Calibration curves for a) acetic acid and b) propionic acid.
$mM = mmol\,l^{-1}$

$$LOD = \bar{x} + 3s \qquad (4.1)$$

$\bar{x}$ - average area measured for an analyte

s - standard deviation

4.2.2 Comparison with direct aqueus sample injection into GC-MS

Limits of detection and limits of quantification (LOQ) of the obtained values using the HS-GC-MS seemed relatively high (in comparison to results obtained by Banel (2010), what will be discussed in details later on). Thereofore they were compared with those for direct injection of a standard VFAs solution.

To find the method of the possibly lowest limit of detection for the VFAs, a method without sample vaporization step (eliminating HS) was developed. Liquid samples were directly injected into the GC-MS system. To allow direct injection of a liquid sample there were two capillary columns installed in the GC. The first one was the FFAP 30 m polar column and the second was only slightly polar, VF-5ms with 5 %-phenyl-95 %-dimethylpolysiloxane. They were connected using a 3 m safety column. The first one should retain the water and the second should provide good separation of VFAs. Chosen operating parameters for GC-MS are listed in Table 4.4.

Table 4.4: Operational parameters fort the GC-MS system.

GC - Varian 431C	MS - Varian 210
Sample volume: 2 µl	Ionization mode: EI
Injector temperature: 250 °C	Ionization energy: 70 eV
Split ratio: off (40 s); 1:20	Trap temperature: 180 °C
Temperature program:	Axial modulation: 2.2 V
$80\,°C \xrightarrow{7\,°C/min} 218\,°C$	
Carrier gas: He (5.0)	
Carrier gas flow: 1.2 ml/min	
Total analysis time: 20 min	

The solution consisting of VFAs at a concentration of $10\,\mathrm{mmol\,l^{-1}}$ was appropriately diluted with methanol as a solvent, which is highly volatile and can evaporate in the injector liner, before the sample reaches the GC column. The obtained chromatograms were characterized by good separation of the analytes, as seen in Figure 4.12. The VFAs of a carbon chain consisting of 2 to 7 carbon atoms in a molecule were identified in the NIST library of mass spectra. The used standard solution contained also formic acid, which could not be separated.

To calculate the LOD and LOQ values, the standard solutions of concentrations in the range of $0.1\,\mathrm{mmol\,l^{-1}}$ to $0.5\,\mathrm{mmol\,l^{-1}}$ were used. The MS was operated in the SIM mode (35-100 m/z). The LOD was calculated using the method based on the standard deviation values of the measurements and the slope of the calibration curve ((4.2) and (4.3)). Limit of quantification was found from the dependence shown in (4.4).

$$LOD = 3.3 \cdot \frac{S}{b} \qquad (4.2)$$

b - slope of the line

S - standard deviation, calculated as a standard deviation of the calibration curve x-intercept S_{xy} (4.3)

$$S_{xy} = \sqrt{\frac{1}{n-2}\Sigma(y_i - Y_i)^2} \qquad (4.3)$$

n - number of separate standard solutions measurements

y_i - experimentally determined value

Y_i - value calculated from obtained equation of regression

$$LOQ = 3 \cdot LOD \qquad (4.4)$$

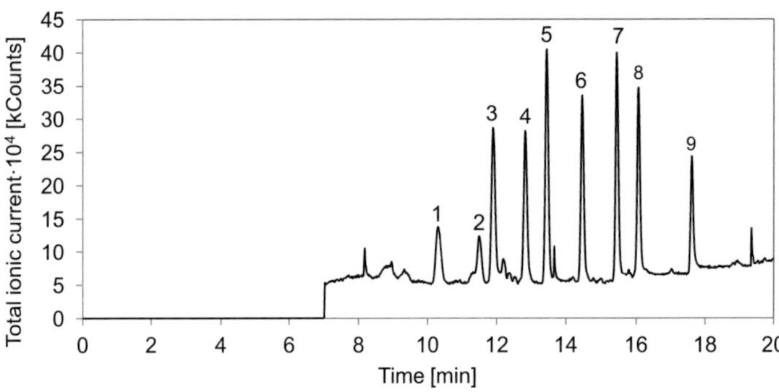

Fig. 4.12: Cinematographic separation of VFAs in the standard solution diluted with MeOH using direct injection: 1- acetic acid, 2- propionic acid, 3- iso-butyric acid, 4- butyric acid, 5- iso-valeric acid, 6-valeric acid, 7- iso-hexanoic acid, 8- hexanoic acid, 9-heptanoic acid; $0.05\ mmol\,l^{-1}$.

The limits of detection and quantification are listed in Table 4.5. The LOD for acetic acid and propionic acid appeared to be ten times lower than those obtained with HS-GC-MS. The main fragmentation ions for acetic acid were

43, 45 and 60; for propionic acid 45, 47 and 74 (Figure 4.13) giving the successful identification and possibility of ^{13}C-labeled analogs differentiation.

Table 4.5: LOD and LOQ values determined for VFAs during direct injection.

VFA	LOD [mmol l^{-1}]	LOQ [mmol l^{-1}]
Acetic acid	0.008	0.024
Propionic acid	0.011	0.033
Iso-butyric acid	0.013	0.039
Butyric acid	0.010	0.030
Iso-valeric acid	0.009	0.027
Valeric acid	0.014	0.043
Iso-hexanoic acid	0.009	0.027
Hexanoic acid	0.012	0.036
Heptanoic acid	0.031	0.093

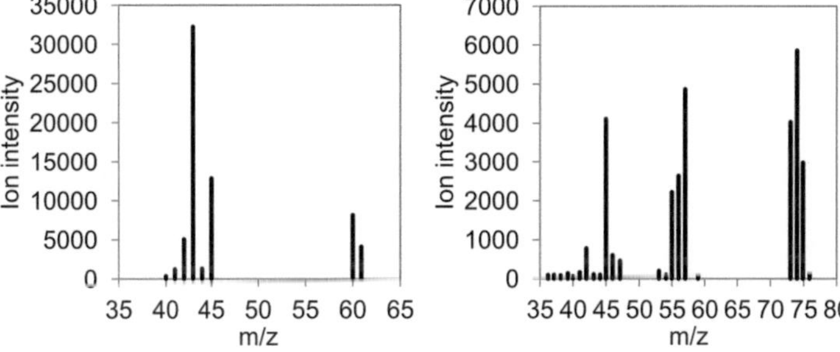

Fig. 4.13: Mass spectra obtained using direct injection into GC-MS for acetic acid (left diagram) and propionic acid (right diagram) dissolved in MeOH.

The next step was the examination of samples with aqueous matrix. Having a better detection limit for acetic acid, the new method was applied for samples taken from anaerobic digesters, fed with ^{13}C-labeled propionic acid. A liquid sample with biomass was first diluted with 50 % (w/ v) $NaHSO_4 \cdot H_2O$ (1:1 ratio) and centrifuged in an Eppendorf cup at 13000 rpm for 2 min. Supernatant was then collected and injected directly into the system. However, probably because of using water as a solvent, the baseline of obtained chromatograms was affected and the ionization failed. The separation of analytes was successful (Figure 4.14), but the differentiation of fragmentation ions was not possible. The m/ z obtained for non-labeled analytes were 43, 46 and 61 for acetic acid and 46, 59 and 75 for propionic acid (compare Figure 3.9 a, b). That denotes that this method is improper for metabolic pathways identification.

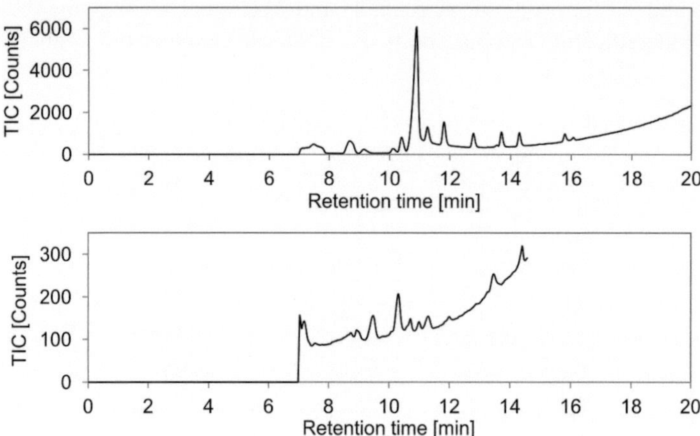

Fig. 4.14: Chromatograms obtained for liquid samples with water as a solvent at different time points: at the beginning of analysis (upper diagram); after several injections (lower diagram).

4.3 Metabolic pathway determination in different habitats

The described HS-GC-MS method was applied for metabolic pathway determination in several mesophilic habitats. The condition that had to be fulfilled in the ecosystem was acetate production during propionic acid degradation. After introducing ^{13}C-labeled propionate, the labeled products were analyzed. Determination of carbon dioxide labeling was impossible, insisting on acetic acid being the major metabolite for pathway determination purpose. The halophilic microbial consortium could not be analyzed, as there was no acetic acid production observed during propionic acid degradation (Figure 4.48). The experimental run from propionic acid feeding to complete degradation and acetate production is represented in Figure 4.15.

4.3.1 Mesophilic biomass from industrial biowaste digester

Organisms originating from industrial scale biowaste digester in Durlach, Germany were analyzed a) directly after collecting them from reactor: "fresh sludge experiment"; and b) after enriching microorganisms on propionic acid as an only carbon source for growth- "enrichment culture experiment".

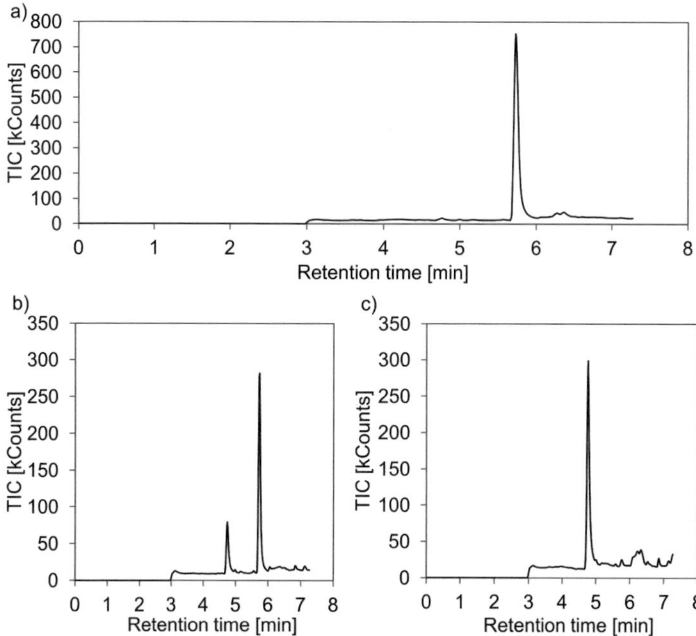

Fig. 4.15: Metabolic pathway determiantion experiment flow from a) propionic acid addition through b) acetic acid production during propionic acid degradation until c) acetic acid formed after total propionic acid degradation.

After $1\text{-}^{13}C$-propionic acid addition to the culture, samples were incubated at 37 °C in the darkness, at pH maintained constantly at 7.0. Methane production, together with propionic acid degradation and acetic acid excretion was controlled at certain time periods until the whole propionic acid was consumed by organisms.

Fresh sludge experiment

The sample obtained from anaerobic digester was first checked for its propionate-degrading activity by adding 2 mmol l⁻¹ propionic acid to 40 ml neutralized biomass suspension (pH 7) in a 100 ml closed serum bottle (performed in duplicate). Propionic acid was degraded within 24 h in both samples and no acetic acid was produced, indicating conversion into methane without measurable intermediate acetate concentration. The surplus methane production, measured to be 8 mmol l⁻¹ and 7 mmol l⁻¹, could be explained by the

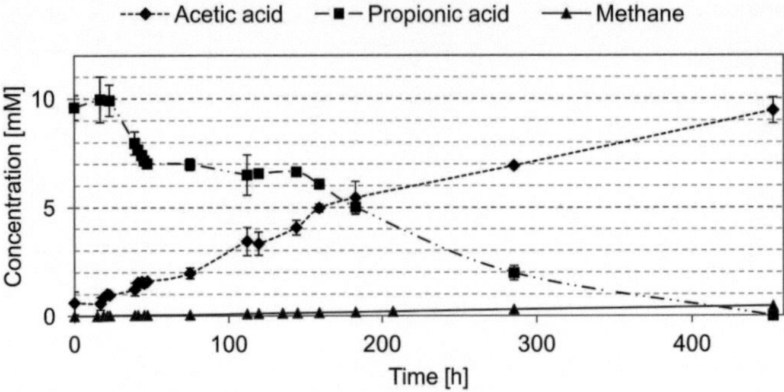

Fig. 4.16: 1-^{13}C- labeled propionic acid degradation by the fresh sludge collected from an industrial-scale biowaste digester. mM = mmol l^{-1}

presence of other organic matter in the samples. Nevertheless, this confirmed that propionic acid degraders were present in the samples and were active. For pathway analysis, 10 mmol l^{-1} BESA inhibiting agent for methanogenesis was added, so that eventual acetic acid produced from ^{13}C-labeled propionate could not be degraded by methanogens, and could be detected by MS. Five ml biomass from digester was introduced into 25 ml test tubes (total volume). From 9.6 ± 0.6 mmol l^{-1} propionic acid, 9.5 ± 0.6 mmol l^{-1} acetic acid was formed within 453 h (18.9 d). The propionic acid degradation rate was 0.03 ± 0.01 mmol l^{-1} h^{-1}. Although the metanogenic bacteria were inhbited, a slight production of methane was also recorded (up to 0.45 ± 0.02 mmol l^{-1}). Overall propionic acid degradation is presented in Figure 4.16. The mass spectrum of the produced acetic acid (Figure 4.17) did not indicate ^{13}C-labeled acetic acid production, eliminating the C-6 dismutation-pathway of propionic acid oxidation as a dominant pathway.

Control experiments with no BESA addition were performed, where 6.38 ± 0.19 mmol l^{-1} propionate was degraded within 49 h. The degradation rate of propionic acid was 0.15 ± 0.07 mmol l^{-1} h^{-1}. In this batch experiment acetic acid was immediately converted to methane. The uninhibited methanogenic partners acted as a catalyst for the propionic acid oxidation reaction in comparison to the reaction in the approach with BESA addition described above. The degradation rate was 5 times faster.

In the sample collected from the reactor, the Smi SR 16S rRNA-based oligonucleotide probe (*Smithella propionica* and *Syntrophus* specific probe) gave a positive signal (Figure 4.18). That could indicate the presence of *S. propioni-*

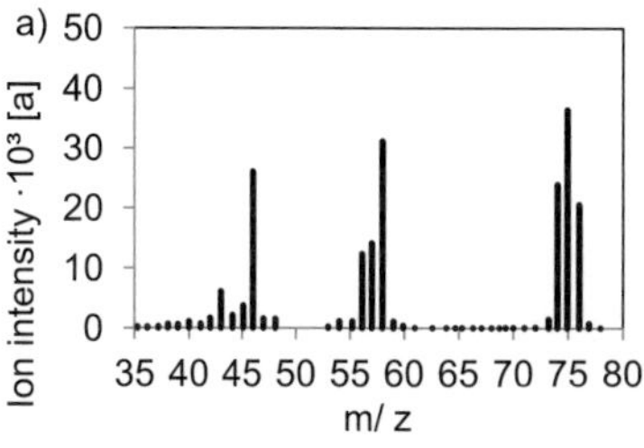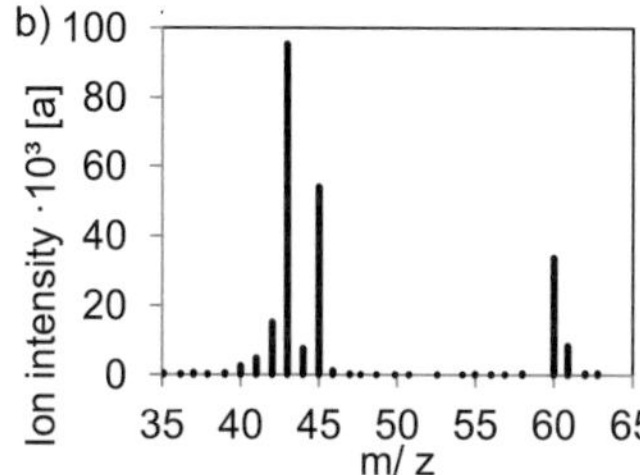

Fig. 4.17: Mass spectra of a) 1-^{13}C-labeled propionic acid supplied as a substrate for "fresh sludge" and b) produced acetic acid with 60 m/ z molecular ion.

ca; however the activity of its metabolic pathway (C-6-dismutation) was not confirmed by ^{13}C-labeled propionic acid experiment.

Enrichment culture

40 ml mixed culture grown after 4 feedings of propionic acid (5 mmol l^{-1} per each feeding) was prepared as described in the section 4.1.1. The mixed bacteria were fed with $\approx$ 5 mmol l^{-1} 1-^{13}C-labeled propionic acid and incubated until the substrate was completely degraded. This procedure was repeted 5 times in order to identify the dominating pathway of propionic acid oxidation. A typical experimental run for an enriched culture is presented in Figure 4.19. The overall balance was 6.63 ± 0.31 mmol l^{-1} acetic acid and 5.52 ± 0.10 mmol l^{-1} methane production from 6.68 ± 0.09 mmol l^{-1} propionic acid within 102 h. The numbers express almost ideally the assumption that 1 mol propionic acid gives 1 mol acetic acid and 0.75 mol methane. Parallel control experiment (no propionic acid addition) produced 0.26 mmol l^{-1} methane, what could explain the surplus methane produced by the culture fed with propionic acid. The propionic acid degradation rate was 0.09 ± 0.04 mmol l^{-1} h^{-1}. There was no record of ^{13}C-labeled acetic acid production during each phase of the experiment, indicating methyl-malonyl-CoA degradation pathway used by the culture. The mass spectra for propionic acid after the feeding and for acetic acid resulting from its oxidation are shown in Figure 4.20.

FISH experiments indicated no members of *Syntrophus* group or *Smithella propionica* presence (see section 4.5).

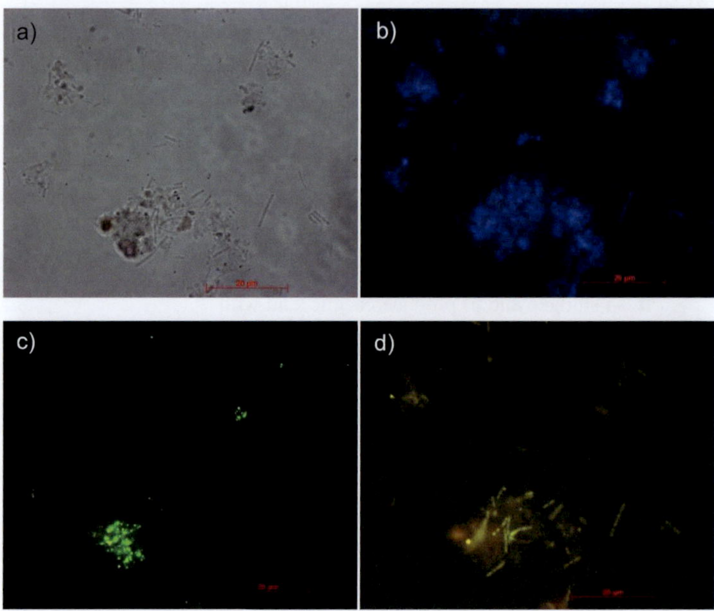

Fig. 4.18: FISH microphotographs of the sample collected from an industrial scale CSTR a) bright field, b) after DAPI staining and after hybridization with c) FAM- labeled Smi SR 354, c) Cy3 labeled Arc 915 oligonucleotide probes.

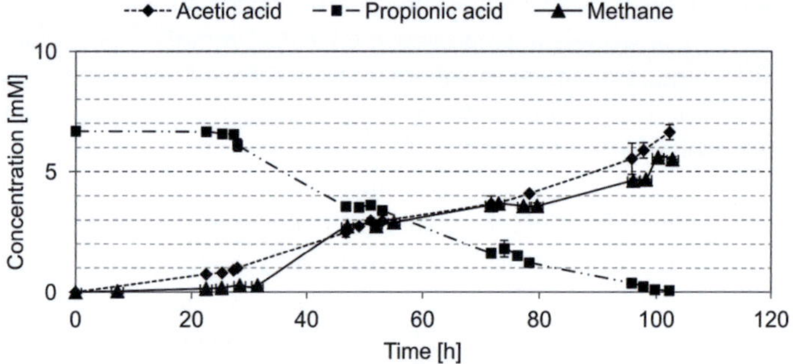

Fig. 4.19: 1-^{13}C-labeled propionic acid degradation by the mesophilic enrichment culture from industrial-scale biowaste digester.
mM = mmol l^{-1}

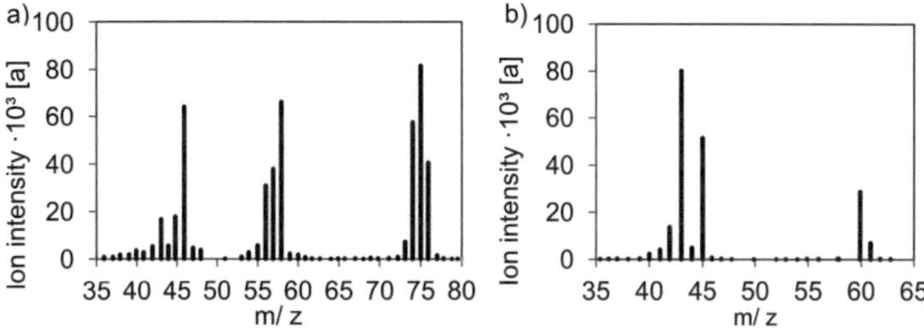

Fig. 4.20: Mass spectra of a) 1-^{13}C-labeled propionic acid supplied as a substrate for the enrichment culture and b) produced acetic acid with 60 m/ z molecular ion.

4.3.2 Biomass from mesophilic lab-scale reactors treating market waste in Indonesia

There were two different lab-scale digesters running in Indonesia at two institutes: LIPI and ITB. Enrichment cultures from both of them, as well as samples originating directly from reactors (sample list containing 5 different samples from ITB and 1 from LIPI is shown in Table 4.6) were examined to determine the active propionate-degrading pathway. Before that, the VFAs concentration was analyzed in the samples 1-6. High levels (except sample 2 and 5) of VFAs were detected in each of them. In order to prepare the biomass suspension for the experiment, the pH was corrected with NaOH to 7.0, and samples were incubated at 37 °C until no residual propionic acid could be recorded. Examined samples are depicted in Figure 4.21.

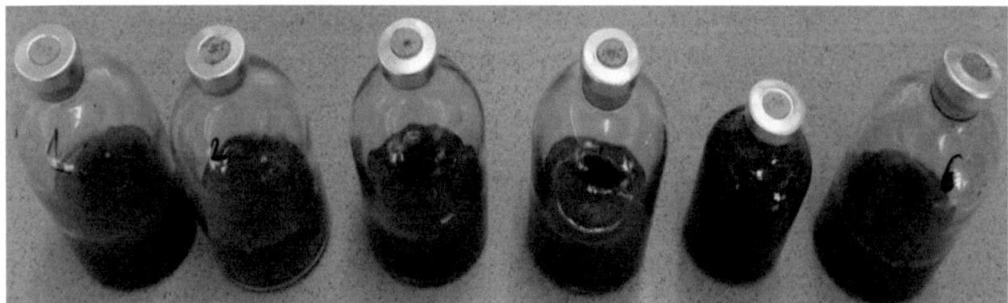

Fig. 4.21: Representation of biomass samples collected from different Indonesian lab-scale reactors treating market waste; from left: samples numbered from 1 to 6, labeling corresponds with description in Table 4.6.

Table 4.6: Description of samples originating from Indonesian lab-scale reactors.

No.	Origin	Description	Initial VFA concentration		
			Ac[1] [mM]	Prop[2] [mM]	Other[3,4]
1	ITB Slurry fraction of reactor from 2nd running	Green-brownish biomass suspension; high turbidity; intensive odour	208.52 ± 14.18	54.10 ± 4.00	n-, i-but[3]: (<90 mM); n-, i-val[4]: (<30 mM)
2	ITB Liquid fraction of continuous fixed-bed reactor	Dark pink/ redish biomass suspension; low turbidity; less intensive odour	1.88 ± 0.13	0.73 ± 0.05	n-but[3]: (¡ 40 mM)
3	ITB Liquid fraction from 2nd running	Brownish biomass suspension; very high turbidity; intensive odour	82.72 ± 5.62	20.81 ± 1.54	n-, i-but[3]: (< 30 mM); n-val[4]: (< 6 mM)
4	ITB Liquid fraction from batch reactor 2nd running	Brownish biomass suspension; very high turbidity; intensive odour	71.17 ± 4.84	30.49 ± 2.26	n-, i-but[3]: (< 23 mM); n-val[4]: (< 6 mM)
5	ITB Liquid fraction from fixed-bed reactor	Red biomass suspension; very low turbidity; slight odour	3.35 ± 0.23	5.57 ± 0.41	none
6	LIPI -	Dark-blackish biomass suspension; very high turbidity; intensive odour	199.68 ± 13.58	13.81 ± 1.02	n-, i-but[3]: (< 10 mM); n-val[4]: (< 5 mM)

[1] acetic acid
[2] propionic acid
mM = mmol l^{-1}

[3] i-but: isobutanoic acid; n-but: n-butanoic acid
[4] i-val: isovaleric acid; n-val: n-valeric acid

Enrichment culture

After 31 days of samples incubation at 37 °C, pH = 7 under a nitrogen atmosphere almost no change in VFAs concentration was noticed. Number 2 (ITB) was the only sample, where total degradation of VFAs was recorded within this time. To eliminate acids from all remaining samples, the centrifuged biomass pellets were washed 3 times with a mineral medium for propionate degraders and suspended in freshly prepared broth (40 ml) in serum bottles (110 ml whole volume). To find out if the propionate degraders were active, 5 mmol l^{-1} propionate was added to each sample and the time for its degradation was measured. The fastest propionate degradation was observed in the sample number 3 (ITB). It took 24 days. Organisms from samples numbered 1 and 6 (Table 4.6) degraded 5 mmol l^{-1} propionic acid within 33 days. Slight or no degradation of the substrate was observed in the samples 2, 4 and 5 after 90 days incubation under constant pH (7) at 37 °C in the darkness. The activity of propionate degraders was evaluated as very low and the samples were not further examined. The main focus was put on the samples of different origin: sample number 3 (ITB) and sample number 6 (LIPI).

The ITB originating sample (no. 3) was prepared for ^{13}C-propionic acid experiment by suspending bacteria in 10 ml broth and feeding of a 2.72 ± 0.02 mmol l^{-1} 1-^{13}C-labeled propionic acid. Duplicate analyses of propionic and acetic acids and methane during incubation for a time period of 94 h were performed (Figure 4.22 a). After 4 days, propionate was almost completely degraded but little less acetate and little more methane than stoichiometrically expected for the methyl-malonyl-CoA pathway of propionate degradation was found. No ^{13}C labeled acetic acid was recorded during the experiment (Figure 4.22 b), excluding the C-6-dismutation pathway.

The mass balance for the experiment is represented in Table 4.7. As 3 mol of hydrogen should have been produced from one mol propionate during anaerobic oxidation (1.1), it appears from (1.3), that 0.75 mol of methane might be produced per mol propionic acid by hydrogenotrophic methane bacteria. In addition, one mol acetic acid should have been produced per one mol propionic acid (Figure 3.7). However, little less acetate was obtained and little more methane than expected was found. Apparently, aceticlastic methanogens that were present in the inoculum converted some of the acetic acid to methane (1.2). Taking all the interactions into consideration, the theoretically expected methane production was 25.62 µmol if 24.67 µmol propionic acid was degraded and only 17.55 µmol acetate left after 94 h (Table 4.7). However, 28.2 µmol methane was obtained, about 10 % more than theoretically expected. It appears from this comparison that conversions have been measured reliably, as 10 % deviation with a mixed culture under the

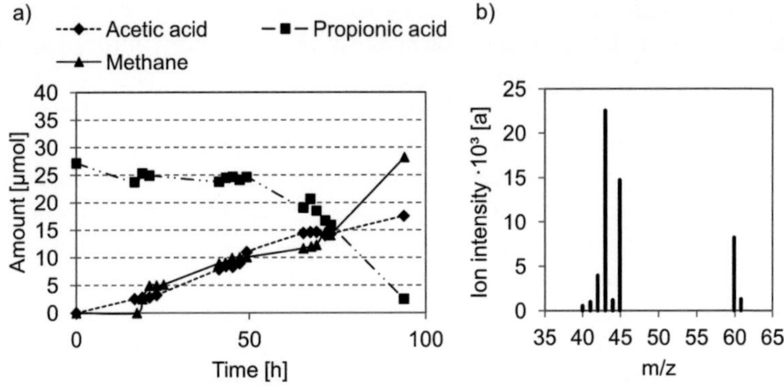

Fig. 4.22: a) 1-^{13}C- labeled propionic acid degradation by the mixed culture from ITB lab-scale reactor treating market waste with b) acetic acid mass spectrum determined after 94 h of incubation.

experimental conditions should generally be acceptable. Some acetate might have also been used for cell proliferation, whereas reserve material might have been fermented to CO_2 and methane by syntrophic interaction at low volatile fatty acid concentrations.

Table 4.7: Mass balance for the 1-^{13}C-labeled propionic acid degradation experiment by the ITB-originating culture; experimental results vs. theoretical considerations.

Amounts [µmol total]		Initially	Measured after 94 h	Theoretically expected[1]
Substrate:	Propionate	27.18 ± 0.21	2.51 ± 0.21	24.67
Products:	Acetate	b. d. l.[2]	17.55 ± 0.21	24.67
	Methane	b. d. l.[2]	28.20 ± 0.85	
Methane:	From CO_2 and H_2[1]			18.50
	From acetate[1]			7.12
	Totally[1]			25.62[3]
	Surplus methane			2.58

[1] if 1 mol propionate is degraded to 1 mol acetate, 1 mol CO_2 and 3 mol H_2

[2] below detection limit

[3] up to 6 µmol methane were produced if the inoculum was incubated in parallel under starvation without propionate addition (data not shown); since propionate was present in our experiment until 94 h we assume that less "background methane generation" occurred from fermentation of intermediates of cell-internal substances

Sample 6 (LIPI) was prepared for the 1-^{13}C-labeled propionic acid experiment according to Figure 4.1, suspended in 5 ml fresh broth and fed with ^{13}C labeled substrate. The concentration of 5.27 ± 0.07 mmol l^{-1} propionic acid was degraded within 50 h resulting in production of 4.93 ± 0.30 mmol l^{-1} acetic acid and 4.13 ± 0.14 mmol l^{-1} methane. Production of a slight methane surplus could be explained analogically to the previously described case. During the control experiment prepared in a fresh medium with no substrate (biomass only), the production of 0.10 mmol l^{-1} methane was measured after 50 h incubation time. Produced acetic acid was identified as non-labeled acetic acid with 60 m/ z molecular ion, indicating the methyl-malonyl-CoA pathway used by the culture.

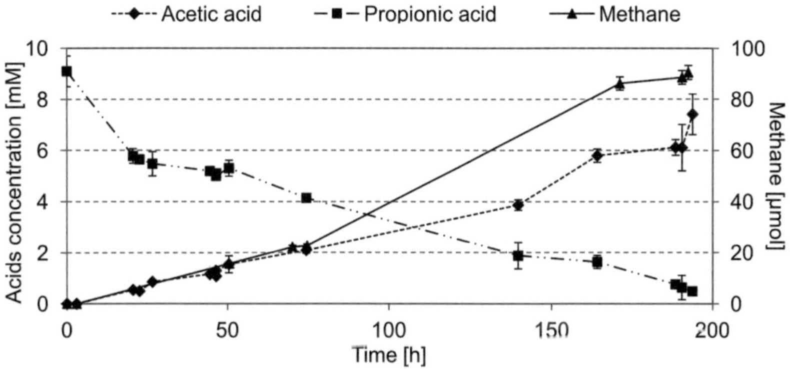

Fig. 4.23: 1-^{13}C-labeled propionic acid degradation by enrichment culture from LIPI lab-sclae market waste digester. mM = mmol l^{-1}

LIPI- originating enrichment sample
The sample for the 1-^{13}C-labeled propionic acid experiment was prepared from 40 ml previously cultivated culture from a mesophilic lab-scale market waste reactor on propionic acid (for 1 month). After washing the biomass with fresh medium and suspending it in 10 ml newly prepared mineral broth, the ^{13}C-labeled propionic acid was added and the experiment began. Degradation of 91.00 ± 6.19 µmol (9.10 ± 0.60 mmol l^{-1}) propionate lasted almost 200 h and resulted with production of 63.77 ± 4.46 µmol (7.41 ± 0.80 mmol l^{-1}) acetate and 90.60 ± 2.71 µmol methane (Figure 4.23). At the end of the experiment the remaining propionate concentration was 4.24 ± 0.39 µmol (0.49 ± 0.08 mmol l^{-1}). Considering the stoichiometry of conversion, the amount of substrate should give 65.07 µmol CH$_4$ and 86.76 µmol acetate.

Calculating the mass balance according to (1.1) and Figure 3.7, 2.54 μmol surplus methane production was measured. That corresponded with the methane produced by control sample, which was incubated without propionate addition (under starvation) and is within the measurement error. The mass spectra of the acetic acid produced from 1-^{13}C-labeled propionic acid by the examined culture did not show ^{13}C incorporation in any of produced metabolites. Both, substrate and produced metabolite spectra are shown in Figure 4.24.

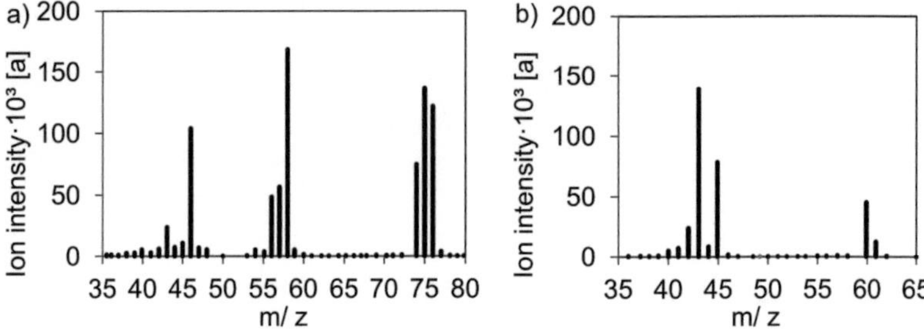

Fig. 4.24: Mass spectra of a) 1-^{13}C-labeled propionic acid supplied as a substrate for the LIPI-enrichment culture and b) produced acetic acid with 60 m/ z molecular ion.

4.4 Microbial isolation and propionic acid degradation rates in different microbial consortia

The basic method of microbial transfer by means of dilution was used for propionic acid degraders' enrichment. For further isolation purposes the dilution row technique was applied together with streak plating on nutrient agar medium and inoculation of a medium with agar addition in roll tubes.

4.4.1 Mesophilic bacteria

Full-scale anaerobic biowaste reactor for biogas production
Microorganisms collected from the biowaste treatment plant in Durlach (anaerobic CSTR), Germany were enriched according to the experimental arran-

gement described in section 3.1.1. Within a 3 years period time transfer of the inoculum was repeated every 30 to 65 days. The change in ecosystem structure was monitored with microscope application. Typical propionic acid degradation curves for these enriched bacteria are shown in Figure 4.25.

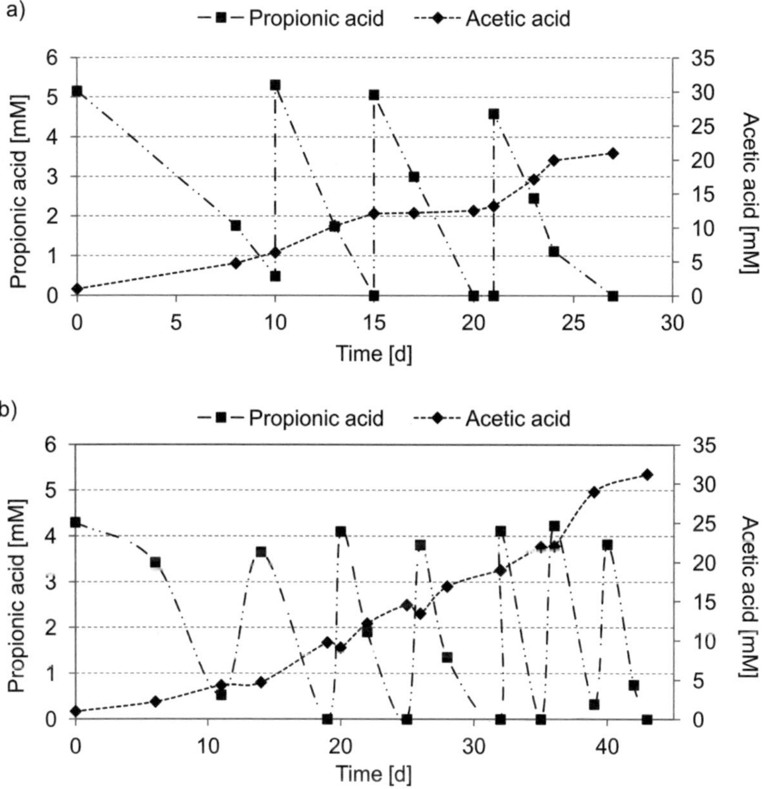

Fig. 4.25: Propionic acid degradation and acetic acid production for the enrichment culture from industrial-scale CSTR: a) early enrichment, b) advanced enrichment. mM = mmol l^{-1}

There was no significant increase in degradation rates for propionic acid within time and increasing enrichment degree. During the first year a trend for increasing propionic acid degradation rate was recorded. The maximum degradation rate measured for a single experiment was 2.83 mmol l^{-1} d^{-1} after the 8th transfer, and the minimal equaled 0.16 mmol l^{-1} d^{-1}. Maximum degradation rates were about 20 times higher than those observed in previous experiments. There were 8 different enrichments for this comparison taken, numbered from 1 to 8 according to the number of microbial transfer (inoculation). Differences in the incubation time and the "advancement" of the enrichment are the basic features for their differentiation. Averaged maximum degradation rates obtained for experiments performed in triplicates are summarized in Figure 4.26. All numbers denote the maximum degradation rates during growth on propionate. In all enrichments acetic acid was produced stoichiometriclally from propionic acid following the assumption that form 1 mol propionic acid 1 mol acetic acid is produced (Figure 3.7).

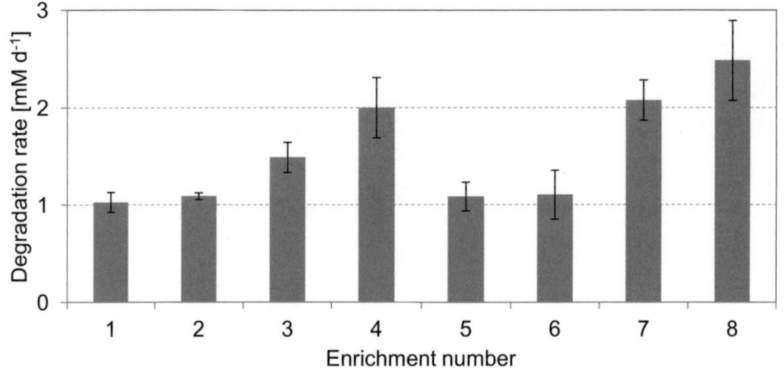

Fig. 4.26: Maximum propionic acid degradation rates for enrichment cultures 1 to 8 from industrial-scale biowaste digester; the number of enrichment describes the transfer number of microorganisms. mM = mmol$\,$l^{-1}

One stage, two phase wet and dry laboratory-scale fermentation plant (ITB)

Enrichments from the culture originating from lab-scale one stage, two phase wet and dry fermentation plant were arranged as described in section 3.1.1. A typical characterization of propionic acid degradation by the microorganisms mixed culture is represented in Figure 4.27. Acetic acid was not produced stoichiomerically from propionic acid (1.1), as in case of industrial scale originating culture. There was no acetic acid production observed within the first 23 days of bacterial growth and a maximum production of almost 5 mmol$\,$l^{-1} acetic acid from ca. 15 mmol$\,$l^{-1} propionic acid was recorded after 50 days of sample incubation. This indicated the presence of acetic acid degraders in the sample.

76

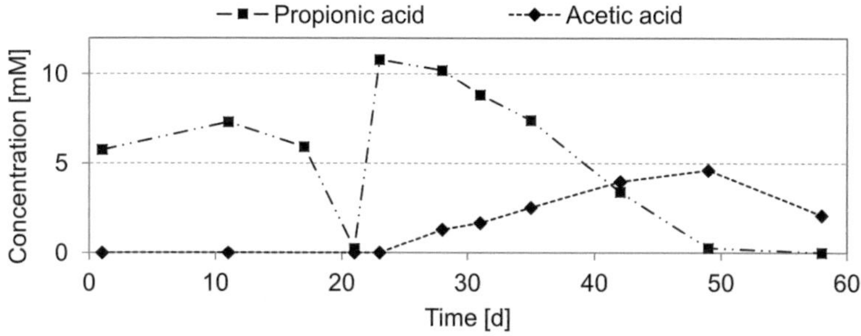

Fig. 4.27: Maximum propionic acid degradation rates for enrichment cultures 1 to 8 from industrial-scale biowaste digester; the number of enrichment describes the transfer number of microorganisms.
mM = mmol l^{-1}

The degradation rates of substrate had values from 0.38 mmol l^{-1} d^{-1} up to maximally 1.37 mmol l^{-1} d^{-1}. Four experiments were compared, and the enrichments were numbered from 1 to 4, where the numbers indicate the enrichment advancement. Because of significantly lower substrate degradation rates, this experiment could be prepared with a number of four subsequent enrichments only. The maximum degradation rate of propionic acid in each set up is summarized in Figure 4.28.

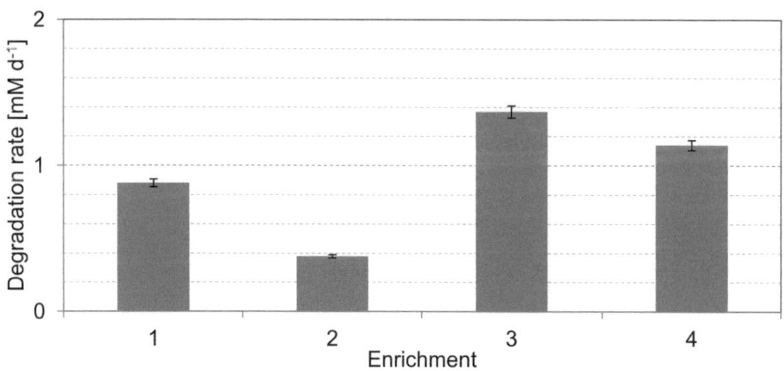

Fig. 4.28: Maximum propionic acid dgeradation rates in enrichment cultures 1 to 4 from lab-scale reactor ITB; the number of enrichment describes the transfer number of microorganisms. mM = mmol l^{-1}

Two stages, one phase laboratory scale wet fermentation plant (LIPI)
The sample with microorganisms from two stages, one phase laboratory scale fermentation plant from LIPI was enriched on propionic acid as described in previous section 3.1.1. The degradation trend of propinate, together with acetic acid production/ consumption is represented in Figure 4.29. From a certain time point on, e.g. after 15 and 38 days, the acetic acid concentration produced from propionic acid degradation started to decrease. This implies that the culture contains acetic acid degraders and that allows organisms oxidation of larger number of propionic acid feedings (because of thermodynamic purposes- elevated acetic acid concentrations inhibit propionic acid degradation, as discussed in section 4.6.2. However, the fact that some medium components are used by microorganisms during their growth, forces the need for their transfer to a freshly prepared broth after a certain time period. The stoichiometry 1 mol acetic acid from 1 mol propionic acid is not maintained due to acetate conversion by aceticlastic methanogens.

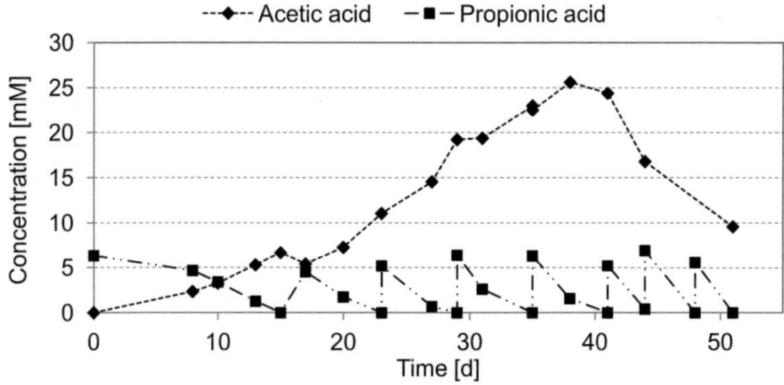

Fig. 4.29: Propionic acid degradation with acetic acid production and degradation by the mixed culture enriched from lab-scale digester LIPI. mM = mmol l^{-1}

The maximum propionic acid degradation rate recorded for one of two parallel running enrichments was 2.93 mmol l^{-1} d^{-1} after 61 days of sample incubation for enrichment no. 8. For the maximum degradation rate comparison the values within 50 days growth were averaged from two parallel runs and compared in Figure 4.30. Maximum degradation rates oscillate in the range from 1.13 mmol l^{-1} d^{-1} to 2.1 mmol l^{-1} d^{-1}.

Differences between mesophilic samples
The samples from German industrial digester and two Indonesian reactors

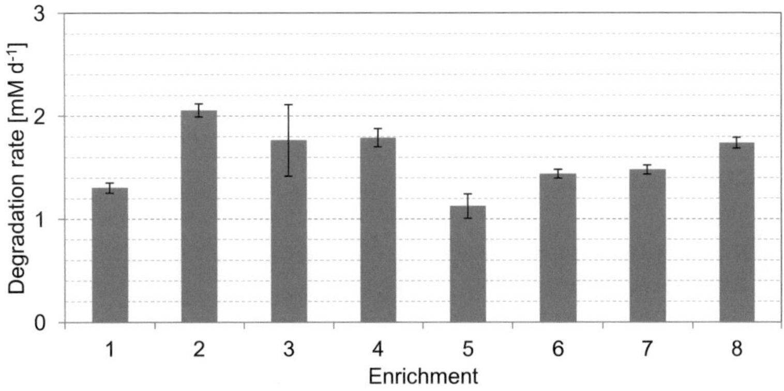

Fig. 4.30: Maximum propionic cid degradation rates for enrichment cultures 1 to 8 from LIPI laboratoryl-sclae biowaste digester; the number of enrichment describes the transfer number of microorganisms into fresh medium. mM = mmol l^{-1}

differ in view of propionic acid degradation rates. Representation of a change in this parameter between first and latest enrichment is shown in Figure 4.33. It seems that rates increased with time, however there was no visible trend of increase for intermediary enrichment numbers in each of the analyzed inocula. Cultures were apparently very sensitive to oxygen and even the low residual oxygen in the anaerobic chamber during centrifugation and preparation of the transfers led to some inactivation with a mixed culture. From the figures representing the degradation trends (Figures 4.25, 4.27 and 4.29) it appears that the ecosystems consisted of different bacteria and this was confirmed by the different morphologies of microorganisms seen in Figure 4.32.

Further approach for enrichment of propionate degraders: supply of different co-substrates

The most active consortium, originating from the industrial scale digester was chosen for further enrichment purposes. The inoculum was distributed into separately prepared fresh media with co-substrates or electron mediators. These were fumaric acid (Figure 4.33 a), crotonic acid (Figure 4.33 b), pyruvic acid and sulfate.

Degradation of propionic acid was recorded in the sample with fumarate (here also succinate was produced from fumarate (Figure 4.33 a), crotonate (Figure

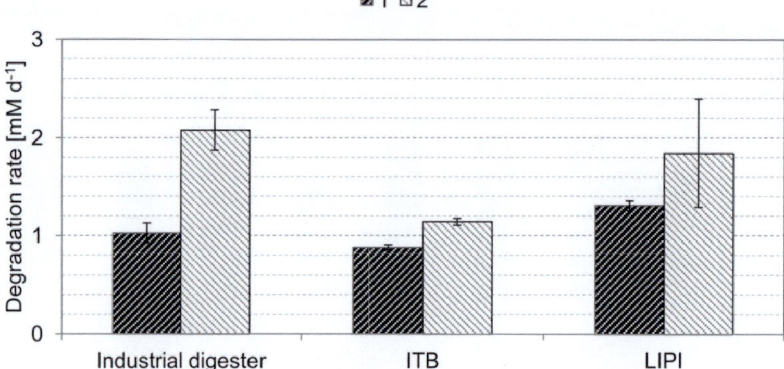

Fig. 4.31: Propionic acid degradation rates change for different microbial mixed cultures in the initial phase (1) and after the 4th (ITB) and 8th (Industrial digester, LIPI) after transfer of the enrichment culture (2). mM = mmol l^{-1}

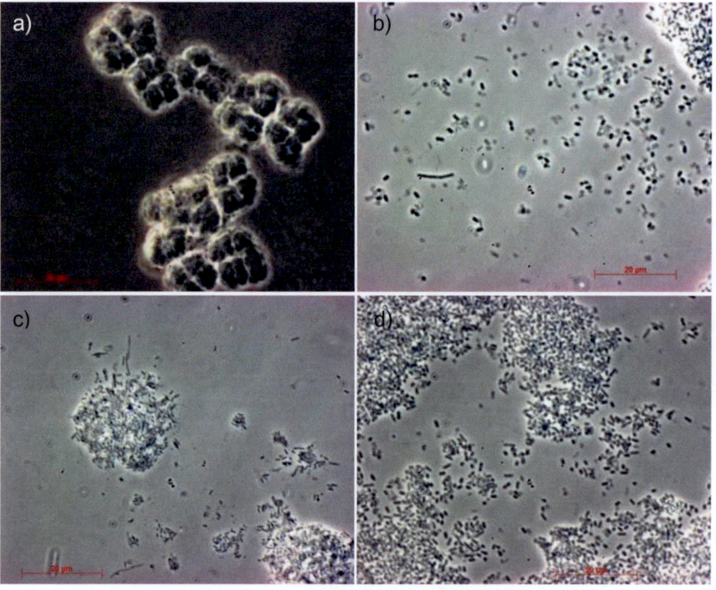

Fig. 4.32: Microscopic differences of the enrichment cultures from anaerobic digesters: a), b) LIPI, c) ITB and d) industrial CSTR.

4.33 b) and sulfate (Figure 4.34) addition. Pyruvate addition caused bacterial growth, however propionic acid did not serve as a substrate and this sample was not further cultivated, as it probably would not lead to enrichment of propionate degraders.

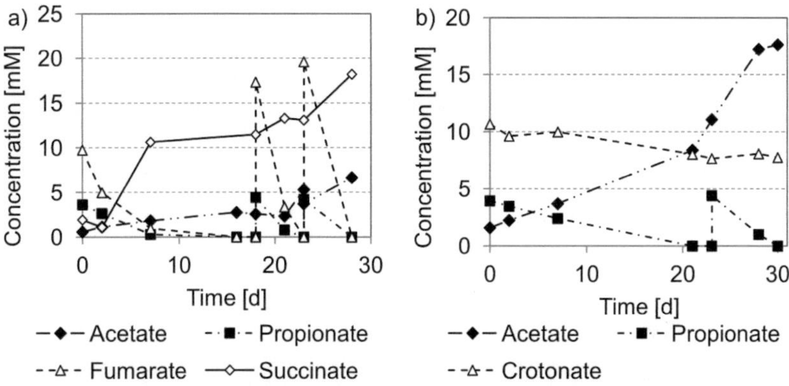

Fig. 4.33: Propionic acid degradation by the mesophilic enrichment culture with co-substrate addition a) fumaric acid and b) crotonic acid.

The most homogeneous sample was obtained after sulfate addition. This sample was pasteurized for 10 minutes at 80 °C to confirm or eliminate the spore forming ability of the organisms. After 2 months of sample incubation there was no degradation of propionic acid observed (no spores formed). The same sample was also distributed into fresh medium with the same co-substrates and BESA as an inhibiting substance for methanogens. In the sample with sulfate $4\,\mathrm{mmol\,l^{-1}}$ propionic acid was degraded after 4 days, whereas the reference sample with the same inoculum and without co-substrate at the same time degraded $1\,\mathrm{mmol\,l^{-1}}$ only.

The enrichment culture obtained after 4 transfers into medium with propionic acid and sulfate has been further enriched by means of dilution series prepared in roll tubes with liquid mineral medium up to 10^{-6}. The highest dilution degree, where degradation of propionic acid was observed, was inoculated into further series. The fastest degradation was observed at the 10^{-5} dilution of second series. It degraded $8.70\,\mathrm{mmol\,l^{-1}}$ propionic acid in two days. This dilution was taken as an inoculum for the roll tube isolation with solidified medium, containing $20\,\mathrm{g}$ agar per liter (shake agar). Black, round colonies were picked in the anaerobic chamber with a sterile syringe and transferred into liquid medium with propionic acid and sulfate. Propionic acid degradation associated with sulfate reduction was recorded. Further dilution series were prepared and the growth of the 10^{-1}- and 10^{-4}- fold dilution is presented in Figure 4.34.

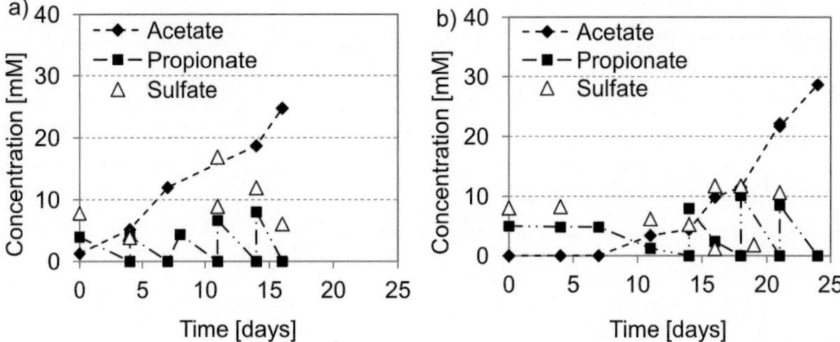

Fig. 4.34: Propionic acid degradation with simultaneous acetic acid production and sulfate reduction for the dilution 10^{-1} (a) and 10^{-4} (b). Second sulfate addition was performed after 8 (a) and 16 days (b) of sample incubation. $mM = mmol \, l^{-1}$

Propionic acid degradation rates for dilution 10^{-1} and 10^{-4} were respectively $4.20 \, mmol \, l^{-1} \, d^{-1}$ and $3.37 \, mmol \, l^{-1} \, d^{-1}$. There were three morphologically different Gram-negative organisms in the sample: lemon-shaped, vibrios and rods (Figure 4.35). In the late growth phase the formation of, most probably, gas vacuoles by the lemon shaped propionate degraders was observed (Figure 4.36). These forms could not be spores, as no culture growth was observed after pasteurization of the sample, and spore staining gave a negative result.

To find out which of the organisms is the propionic acid degrader, isolated enrichments were distributed into glass test tubes with a nitrogen atmosphere and acetic acid; acetic acid and sulfate, sulfate or propionic acid were added. There was also a sample prepared with a hydrogen and carbon dioxide atmosphere without any other substrate additions and a separate sample with H_2/CO_2 and sulfate. No bacterial growth was observed in the samples with nitrogen atmosphere. The biomass growth confirmed by microscopic observation was recorded in the sample with H_2/CO_2 and sulfate only. Vibrios were enriched in this sample.

Kinetics of propionic acid degradation with sulfate reduction were examined (Figure 4.37 a, b). For determining whether the propionate degrader is also a member of the microbial community that can reduce SO_4^{2-}, molybdate as inhibiting substance for sulfate reduction was added to the sample at 16th day of

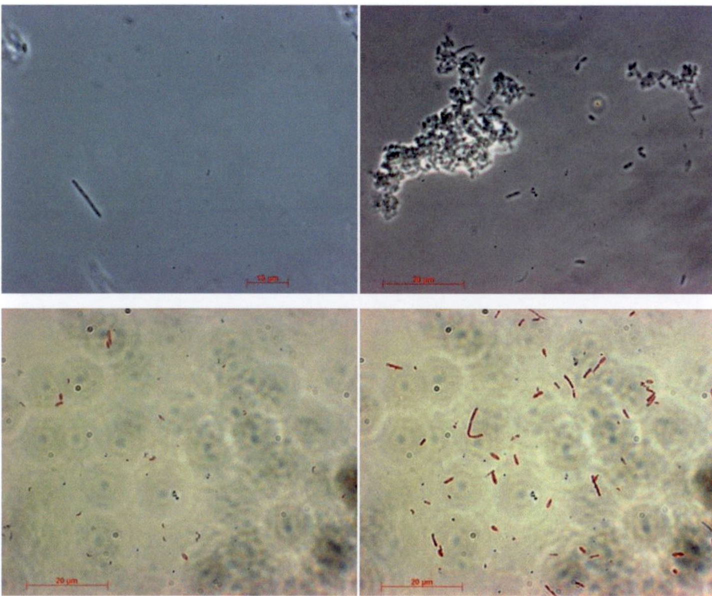

Fig. 4.35: Organisms isolated from the industrial scale anaerobic digester - phase contrast microphotograph (two upper images) and Gram staining (two lower images).

incubation (Figure 4.38). As the sulfate was further reduced by the organisms at a certain rate, the inhibition of sulfate degraders was not complete. However, a significant influence on the rate of anaerobic oxidation of propionic acid was noticed, and hence, described as inhibition of propionate degraders. Elevated concentrations of acetic acid and propionic acid measured after 33rd day are caused probably by the molybdate oxidation state change- the sample changed the color from transparent to brown, most probably indicating the presence of molybdenum (V).

4.4.2 Thermophilic bacteria

Anaerobic sludge from an industrial dry biowaste fermenter in Leonberg was taken and suspended in the mineral medium for propionate degraders. After several transfers of the suspension fed with propionic acid as the sole carbon source, the sample was pasteurized at 80 °C for 10 minutes and fed again with propionic acid. In such a way the spore forming organisms that can degrade

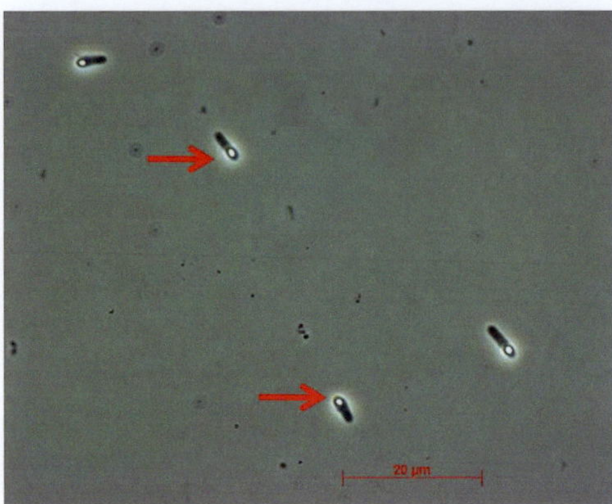

Fig. 4.36: Gas vacuoles (marked with arrows) formed by mesophilic propionate degraders during late growth phase. Spores would result in the same appearance.

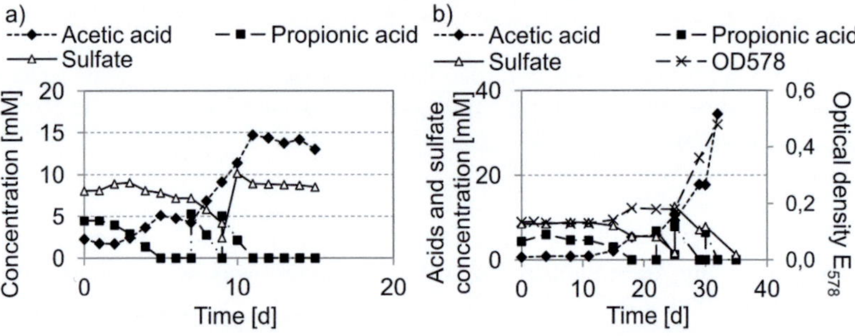

Fig. 4.37: Propionic acid degradation with sulfate reduction and acetate production by the mesophilic enrichment culture, containing three morphologically different bacteria: a) daily concentration measurements of metabolites and sulfate and b) metabolites measurements with additional biomass growth control (optical density). Second addition of sulfate was performed after 10 (a) and 25 days (b) of sample incubation. mM = mmol l^{-1}

propionate could be enriched. Sulfate was added as an electron mediator for hydrogen consumption produced from propionate oxidation, as there are yet no methanogenic bacteria known that are able to produce spores and a low pH$_2$ must be maintained. Degradation of propionic acid was observed, as well as sulfate reduction (Figure 4.40 a). Organims suspected to be the propionic

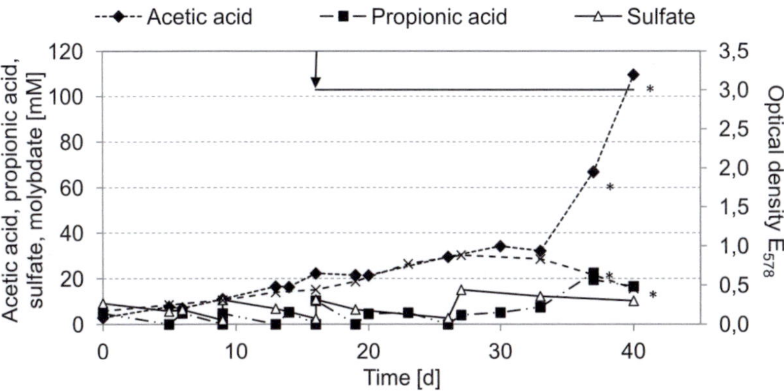

Fig. 4.38: Propionic acid degradation by the mesophilic enrichemnt culture with three morphologically different bacteria. Molybdate addition is marked with an arrow; * denote the measurements influenced by the presence of molybdenum in the sample, considered as higher than in the reality. mM = mmol l^{-1}

acid degraders were morphologically similar to the mesophilic lemon-shaped rods and formed spores (Figure 4.39).

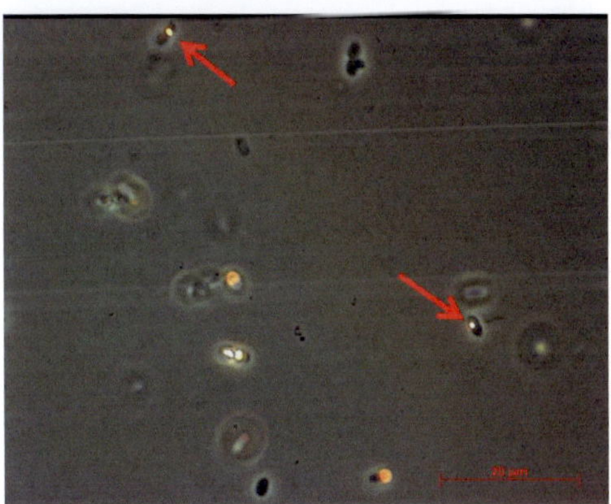

Fig. 4.39: Lemon-shaped thermophilic rods forming spores (pointed with arrows).

Organisms did not grow on sulfate only, and could not degrade acetic acid. However, they grew on propionic acid without sulfate addition, what lead to assumption, that methanogenic bacteria survived the pasteurization (Figure

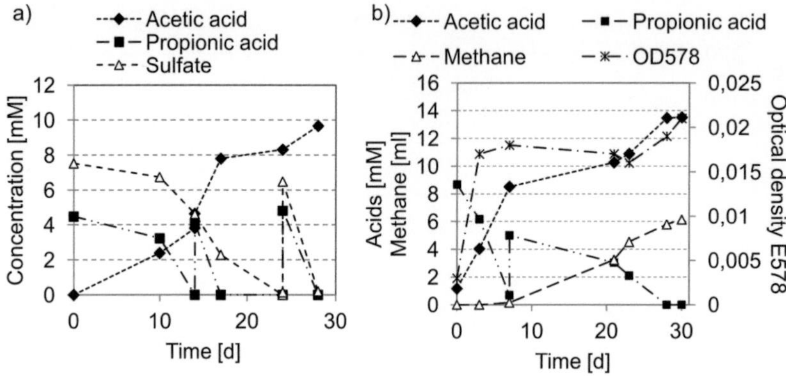

Fig. 4.40: Propionic acid degradation under thermophilic conditions after sample pasteurization a) with sulfate addition, b) no sulfate addition. mM = mmol l^{-1}

4.40 b). This was confirmed with measurement of methane production in the sample with propionic acid only, and in the sample which was fed later with a H_2/ CO_2 gas mixture (Figure 4.41).

Autofluorescence of methane bacteria was observed under the UV light due to the presence of F_{420} enzyme.

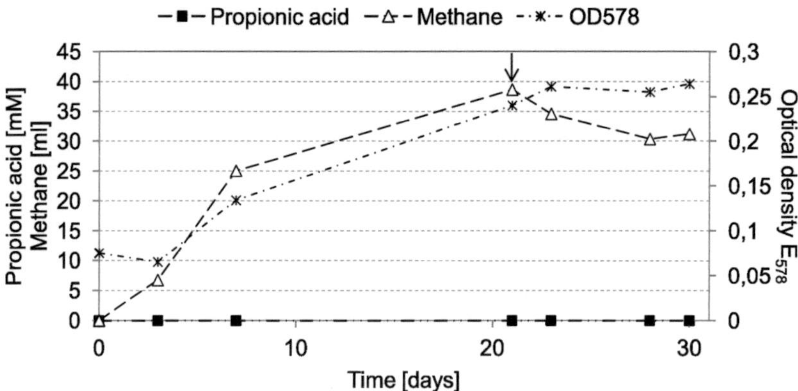

Fig. 4.41: Methane production by the thermophilic methane bacteria from H_2/ CO_2 substrate; the arrow denotes feeding point- gas exchange to H_2/ CO_2 (2 bar). mM = mmol l^{-1}

Steps towards the separation of methanogens and propionic acid degraders' were undertaken, and the isolation of both organisms was arranged. The methane bacteria were cultivated further by means of further enrichment with a H_2/CO_2 gas mixture in dilution rows. A pure culture was accomplished after 5 transfers of inocula, incubated at 70 °C (as no growth was observed at 80 °C). The working name of the strain was MFZ-1. The temperature optimum of the isolated culture was 65 °C (Figure 66). Several substrates were tested for their degradation to methane by the organism: formate, methanol, ethanol, acetate, pyruvate, glucose, lactate, methylamine, trimethylamine. Finally none of these substrates resulted in methane production.

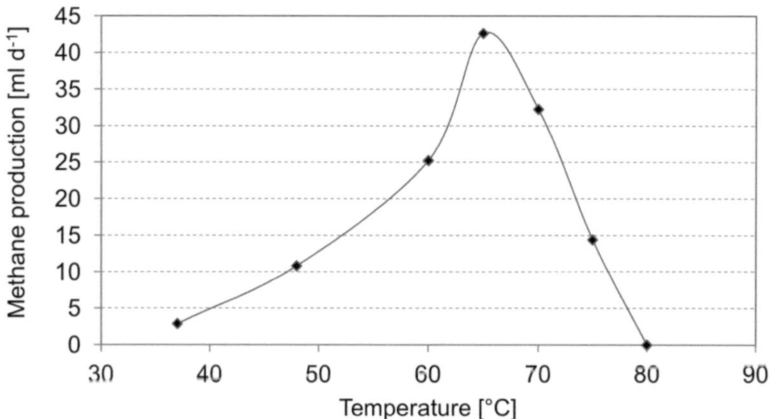

Fig. 4.42: Temperature optimum for methane production by thermophilic methanogen MFZ-1.

The culture where both organisms (propionate degraders and methanogens) were present was inhibited with BESA as a toxic agent for the methane producers, so that the propionic acid degraders could be isolated. 5 mmol l⁻¹ propionate has been degraded within 60 days. Further cultivation was undertaken in order to obtain a pure culture of these bacteria.

A test with the isolated culture to determine if the propionic acid degrader is responsible for sulfate reduction was also performed. The following samples were prepared: with addition of molybdate, BESA and propionic acid; propionic acid, sulfate and BESA, propionic acid and BESA. Results are represented in Figure 4.43.

It appears that the propionic acid degrading bacterium is also responsible for sulfate reduction. In case where both BESA and molybdate were present

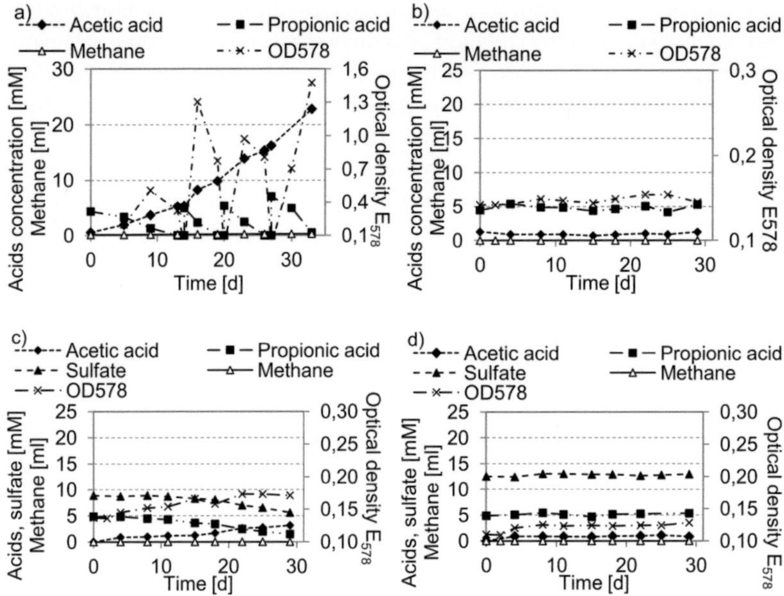

Fig. 4.43: Thermophilic culture isolated from an industrial biowaste digester with a) propionic acid, b) propionic acid and BESA, c) propionic acid, sulfate and BESA, d) propionic acid, sulfate, BESA and molybdate. mM = mmol l^{-1}

(Figure 4.43 d), neither growth was observed, nor methane production or propionic acid/ sulfate degradation were recorded. Whereas in the sample with BESA addition, sulfate was reduced (served as electron acceptor) and propionic acid oxidized (Figure 4.43 c).

The fastest propionic acid degradation rate recorded during the experiment with a culture that contained methanogens was 1.5± 0.11 mmol/l/d. The degradation rate obtained for the culture with methanogens and sulfate present was 1.4±0.1 mmol l^{-1} d^{-1}.

Morphology of isolated bacteria and autofluorescencing methanogens are represented in Figure 4.44.

4.4.3 Rumen liquid

A sample of microorganisms collected from a cow rumen was tested for its ability to degrade propionic acid. Organisms were inoculated into mineral

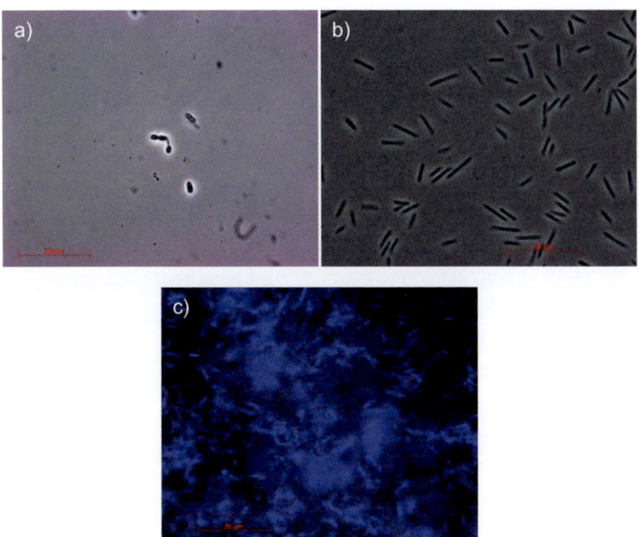

Fig. 4.44: Morphology of isolated thermophilic culture members a) propionic acid degraders, b) methane bacteria during the early growth phase, c) culture under UV light.

medium with propionic acid addition and the oxidation of propionic acid was followed (Figure 4.45).

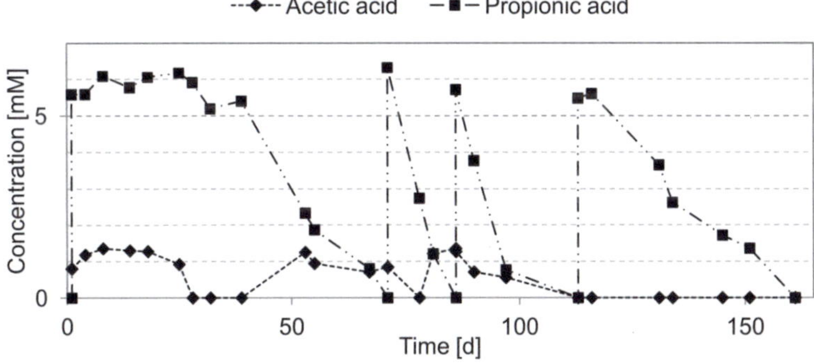

Fig. 4.45: Propionic acid degradation by organisms collected from rumen fluid. mM = mmol l^{-1}

Acetic acid was consumed by the organisms present in the culture. The degradation rate of propionic acid was in the range from $0.14\pm0.07\,\mathrm{mmol\,l^{-1}\,d^{-1}}$ up to $0.45\pm0.01\,\mathrm{mmol\,l^{-1}\,d^{-1}}$. Further growth of microorganisms after their transfer was not successful and degradation of propionic acid was not observed. The morphology of organisms is shown in Figure 4.46. Apparently essential nutrients that were present in rumen fluid were diluted out in the subcultures.

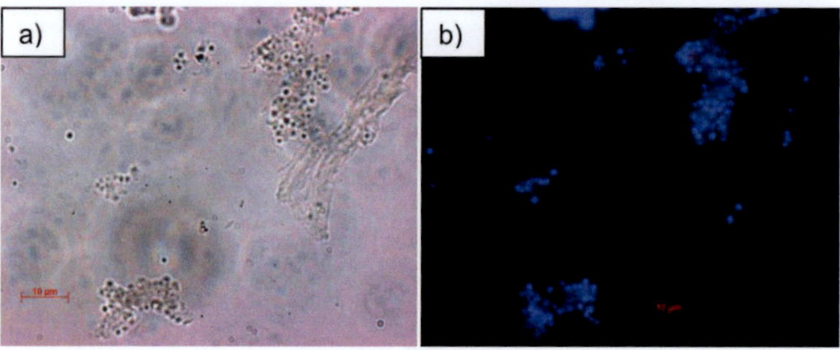

Fig. 4.46: Microscopic images of ruminant bacteria a) bright field and b) after DAPI staining.

4.4.4 Halophilic bacteria

Sediments collected from the North See were also examined for the presence of propionic acid degraders. 4 ml of black sediments were suspended in an anaerobically prepared mineral medium for halophilic bacteria, containing artificial sea water. As within 30 days of incubation at room temperature (20 °C) no microbial growth was observed, the sample was eliminated. Another experiment was arranged with mineral medium for propionic acid degraders with addition of $16\,\mathrm{g\,l^{-1}}$ NaCl. Propionic acid degradation was observed. However, after a certain time period no further oxidation of a propionate was recorded. Methane was measured, and the lack of carbon dioxide in the serum bottle atmosphere has been stated (Figure 4.48). The gas phase was exchanged against N_2/CO_2, what allowed further propionic acid degradation. The culture did not produce any acetic acid. Mineral medium used for culture cultivation was colorless at the time of inoculation. During microbial growth, resazurin was reduced to resofurin (Figure 4.47), the reasons for this are however not clear. Serum bottle headspace analysis did not show oxygen presence in the sample. Additionally, methane production was observed.

Fig. 4.47: Halophilic enrichment culture during growth on propionate- reasazurin color change observation.

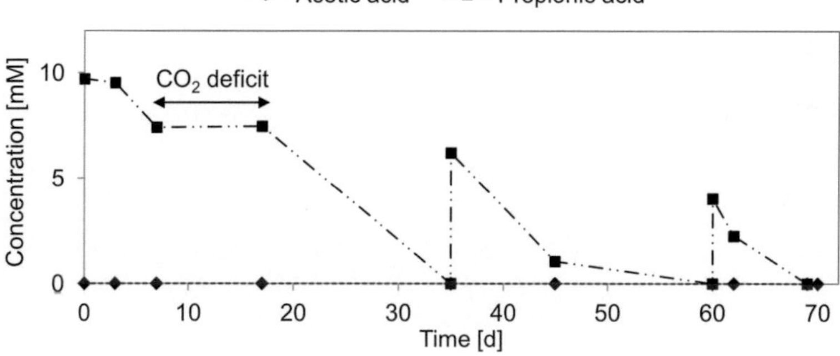

Fig. 4.48: Propionic acid degradation by the halophilic bacteria originating from North See sediments. mM = mmol l^{-1}

Sample contained long rods of methanogenic bacteria, fluorescencing under the UV light and mostly coccoidal other organisms. Aggregation of culture members was observed (Figure 4.49). A confirmation of the presence of methanogens is shown in the Figure 4.50, where a halophilic sample image was taken after hybridization with Cy3-labeled Arc 915 oligonucleotide probe.

The fastest degradation rate of propionic acid by this culture was $0.89\pm0.01\,\text{mmol}\,l^{-1}\,d^{-1}$.

The sample was divided into several different sub-cultures, to isolate propionic acid degraders on a similar basis as previous enrichments of mesophilic and thermophilic bacteria. Sub-cultures with acetic acid, propionic acid and sulfate, as well as H$_2$/ CO$_2$ were arranged. In the sample with acetic acid no

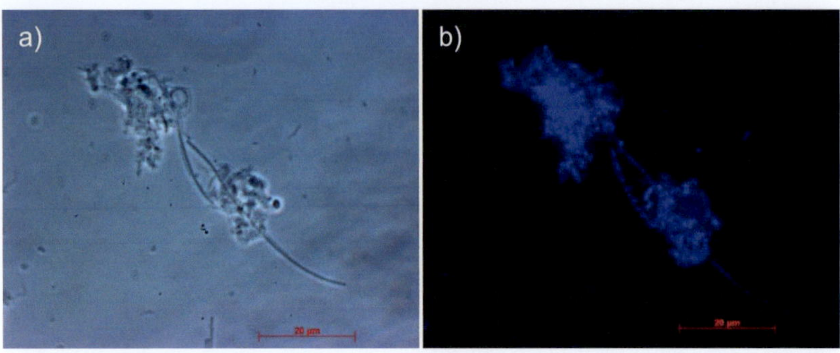

Fig. 4.49: Microscopic images of halophilic bacteria DAPI staining.

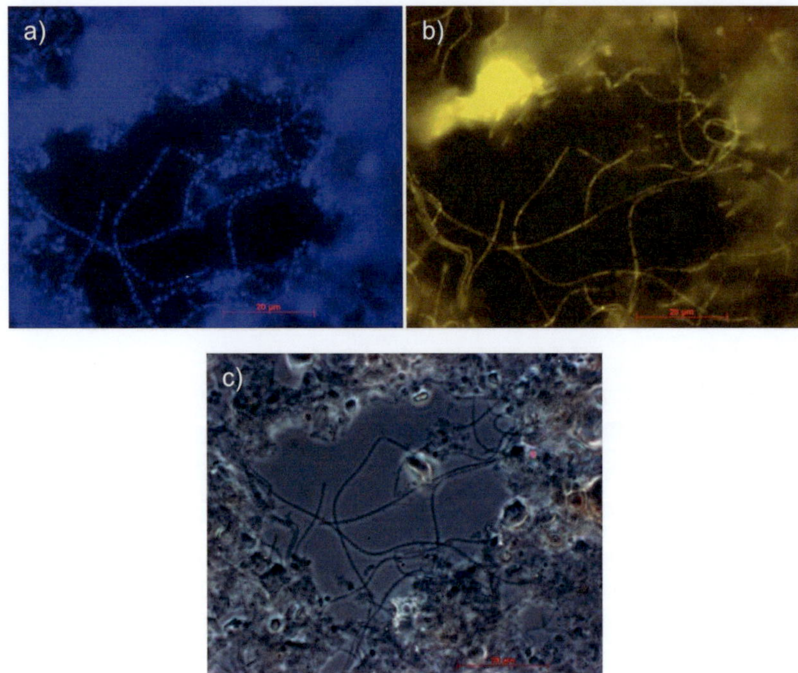

Fig. 4.50: Halophilic bacteria image after a) DAPI staining, b) hybridization with Cy3-labeled Arc 915 oligonucleotide probe, and c) phase contrast microphotograph.

growth was observed within 70 days. Very slow degradation was recorded in the sample with propionic acid as a substrate for growth and sulfate as hydrogen sink. Sulfate reduction during propionate oxidation was measured. $2.16\pm 0.06\,\mathrm{mmol\,l}^{-1}$ sulfate was reduced within 27 days of incubation while $2.86\pm 0.20\,\mathrm{mmol\,l}^{-1}$ propionate was oxidized and some acetate was released $(0.26\pm 0.02\,\mathrm{mmol\,l}^{-1})$. There was no record of methane production by microorganisms in this sample.

4.5 Fluorescence *in-situ* hybridization for microbial identification

Enrichment culture members were identified with the use of fluorescence *in-situ* hybridization. For method validation pure cultures of *Smithella propionica* and *Syntrophobacter fumaroxidans* were analyzed with specific 16S rRNA- based oligonucleotide probes: Smi SR 345 specific for *S. propionica* and *Syntrophus* bacterial group; Delta 495 specific for *Deltaproteobacteria*. Results of hybridization of reference pure cultures are shown in Figure 4.51. Synbac 824 did not hybridize with bacteria in the sample. Syb 701 gave only very weak signal for samples originating directly from CSTR digester and no signal for any other culture examined during this study.

4.5.1 Tri-culture isolated from an industrial scale mesophilic anaerobic biowaste reactor

The enrichment culture of bacterial species isolated form an industrial scale anaerobic reactor contained at least three different bacterial species (according to microscopic observation), where the lemon shaped organisms were identified as Eubacteria and propionic acid degraders belonging to the *Syntrophobacter* group (Figure 4.52) that are capable of sulfate reduction (Figure4.53). The *Vibrios* suspected for sulfate reduction were also classified as SRB after hybridization with SRB 385 oligonucleotide probe.

4.5.2 Thermophilic enrichment culture

The propionate degrader in the thermophilic culture was identified as a member of the *Syntrophobacter* group after hybridization with Synbac 824 16S

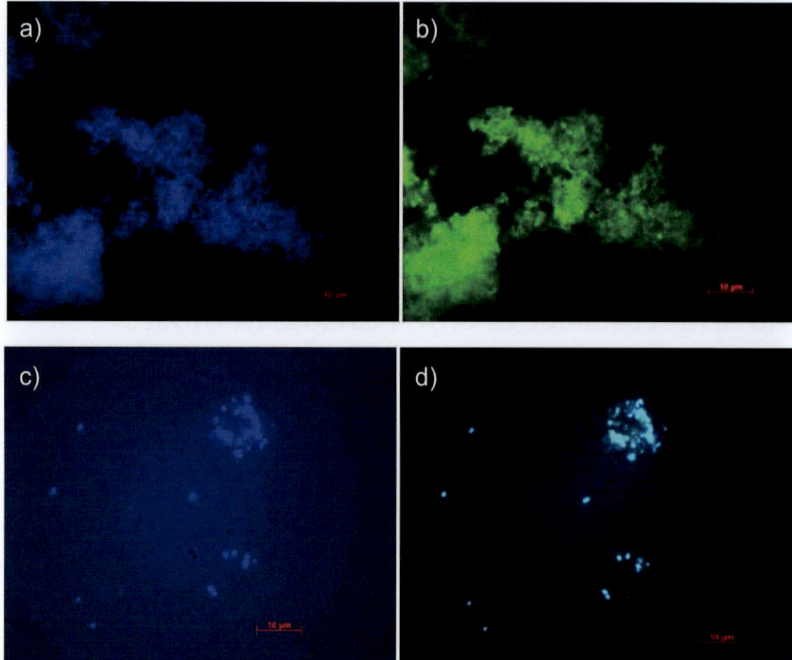

Fig. 4.51: Pure cultures of *Smithella propionica* (a and b) and *Syntrophobacter fumaroxidans* (b and c) after a), c) DAPI staining and b) hybridization with FAM- labeled Smi SR 354 probe, d) hybridization with DY-415- labeled Delta 495 probe.

rRNA- based oligonucleotide probe (Figure 4.54). Organisms did not hybridize with the Syb 701 probe.

4.5.3 Isolated cultures- summary

There were two advanced enrichment cultures of propionic acid degraders and partners isolated. The cultures can be characterized by several features, which are summarized in Table 4.8.

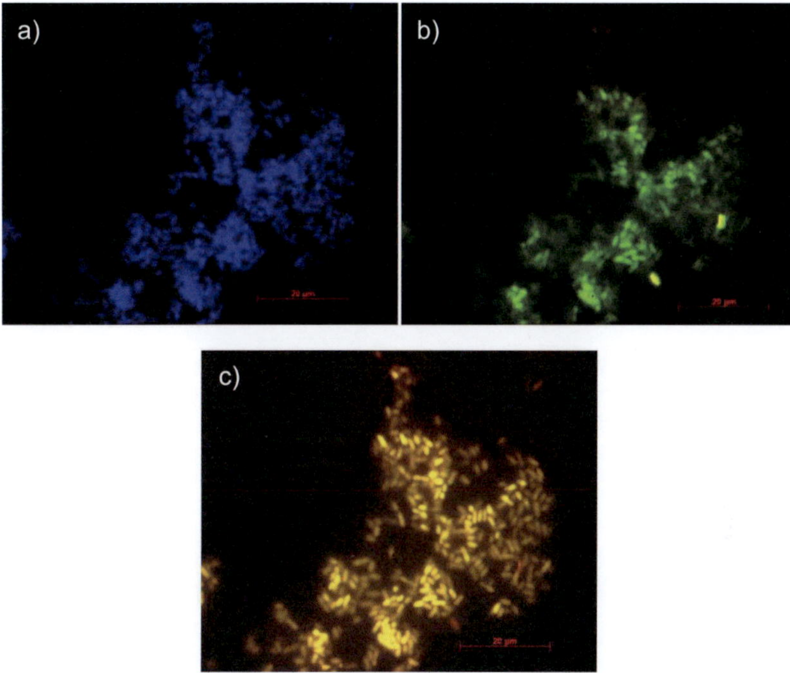

Fig. 4.52: Microscopic images of a tri-culture sample after a) DAPI staining, hybridization with b) FAM- labeled Eub 338 oligonucleotide probe and c) Cy3-labeled Synbac 824 oligonucleotide probe. Bar represents 20 µm.

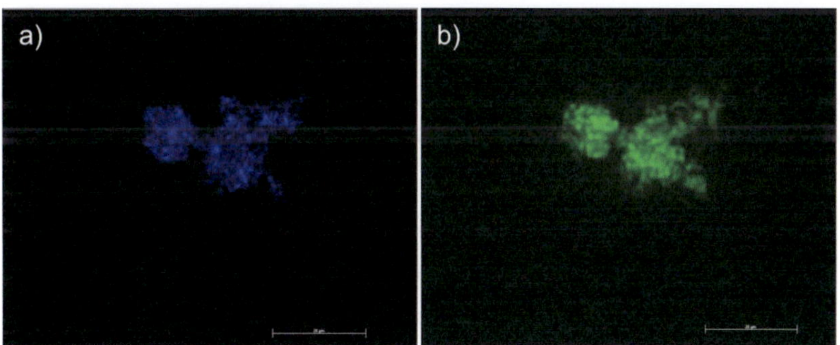

Fig. 4.53: Microscopic images of a tri-culture sample after a) DAPI staining and b) FAM- labeled SRB 385 oligonucleotide probe. Bar represents 20 µm.

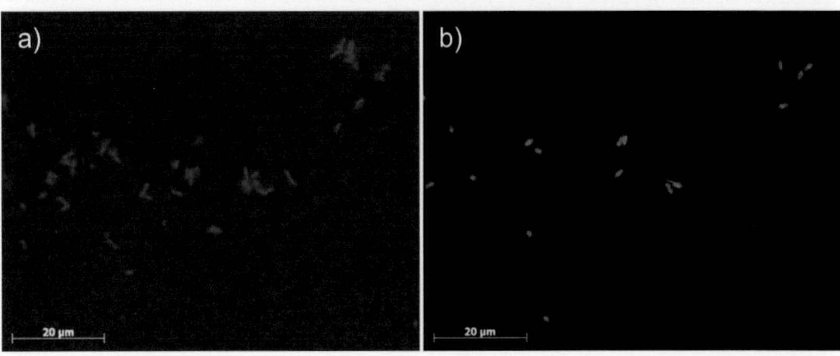

Fig. 4.54: Microscopic representation of a thermophilic enrichment culture after a) DAPI staining and b) hybridization with the Cy3- labeled Synbac 824 oligonucleotide probe. Bar represents 20 μm.

Table 4.8: Features of enrichment cultures.

Feature	Mesophilic "Tri-culture"			Thermophilic culture	
Morphology	Lemon-shaped rod	Vibrio	Rod	Lemon-shaped rod	Rod; forming chains
Size [μm]	1.0-1.6 x 1.2-2.5	0.5-1.0 x 3.0-5.0	0.8-0.9 x 4.4-6.5	0.9-1.4 x 2.4-3.2	0.9-1.0 x 4.2-11.2
Description according to FISH result	*Eubacteria, Syntrophobacter,* MPOB1, SRB	Eubacteria, SRB	*Eubacteria*	*Eubacteria, Syntrophobacter,* SRB	*Archaea* (hydrogenotroph)
Gram staining	G -	G -	G -	G -	G +
Spore formation	-	-	-	+	-
Gas vacuoles	+	-	-	-	-
Methane production		-			+
Molybdate inhibition		+			+
Propionate degradation rate		296 mg l^{-1} d^{-1}			102 mg l^{-1} d^{-1}
Sulfate reduction rate		167 mg l^{-1} d^{-1}			75 mg l^{-1} d^{-1}

4.6 Spatial distances between propionic acid degraders and their partners

Three double experimental runs have been prepared, where each of them was inoculated from the previous one. Samples were fed with $5\,mmol\,l^{-1}$ propionic acid portions, and acetic acid production together with methane formation was monitored. Whenever propionic acid was degraded, a sample of biomass was collected for FISH as a tool for differentiating between both organisms' groups and the measurement of the distances between syntrophic partners. Additionally the aggregate dimensions change within the growth phase was observed.

Several runs differ with a length of a lag phase, degradation time of the same propionic acid concentration, and acetate production. Comparison of the chosen parameters for prepared experiments is presented in Table 4.9. During first two double runs (I and II) samples for FISH analysis were collected at different growth phases and examined after the whole experiment time. The third run (III) was undertaken to do this analysis directly after the sample collection. Here, there was no record of a lag phase. At each of the experimental runs, samples for FISH analysis were collected at the beginning of the experiment, at the end of it and after each propionic acid feeding degradation (the experiment was altogether divided into 5 different "growth" phases, during which the distances were measured). In the run I and II, growth of the culture was monitored with optical density (OD_{578}), however the method was found as inappropriate in case of aggregate formation (see section 3.6). This is why during the third experimental run an assay with protein concentration measurement was performed to receive a better definition of bacterial growth. Furthermore, this assay will be described in detail for determining the interspecies distances change during the culture growth and its influence on propionic acid degradation.

The propionate-degrading enrichment of bacteria with protein concentration measurement was cultured in 250 ml Schott bottle. One hundred ml mineral medium were inoculated with 10 ml of a suspension from the second experimental run. Directly after inoculation and every day later on, 1 ml samples were withdrawn for analyses of propionic acid, acetic acid and the protein content (as a measure of cell growth). Methane, carbon dioxide and hydrogen were withdrawn from the gas phase and determined every day by gas chromatography. Whenever the portion of propionic acid was degraded, biomass was collected for fluorescence *in-situ* hybridization in order to differentiate between propionate degraders and hydrogenotrophic methane bacteria. Or-

Table 4.9: Three double experimental runs with propionic acid for interspecies distances measurement.

Experiment run	I		II		III	
Lag phase [d]	30	35	14	14	None	None
Total incubation time [d]	62	65	50	50	33	42
Hybridization	After whole experiment time				Directly aftersample collection	
OD_{578} increase[1]	0.193	0.116	0.116	0.243	0.154	$(75.8^{2)})$
Propionic acid degraded [mM]	39.54	42.31	20.59	32.61	21.69	20.30
Acetate produced [mM]	33.87	35.80	19.77	27.78	19.50	20.80
Methane produced [mM]	29.50	31.22	15.66	24.87	17.86	15.50

[1] Difference in the extinction value (E_{578}) measured at the time point of inoculation and on the last day of sample incubation;

[2] mgl^{-1} protein (concentration difference between values measured on the first and the last day of sample incubation);

$mM = mmol l^{-1}$

ganisms were identified after reaction with 16S rRNA-based oligonucleotide probes specific for Bacteria, propionate degraders and Archaea. The experiment was divided into 5 phases, namely phase 0 (directly after inoculation), growth phase 1 for degradation of the 1st portion of propionate (until day 13), growth phase 2 after the 2nd propionate feeding (day 22), growth phase 3 after the 3rd propionate feeding (day 29) and growth phase 4 after the 4th propionate feeding (day 42). The experimental run is shown in Figure 4.55.

It was difficult to collect homogeneous samples as the microorganisms formed aggregates. This is the reason why the protein curve was not as smooth as the curves for methane and acetate (Figure 4.55). Nevertheless the increase of the protein concentration revealed a typical exponential behavior over time resulting in a doubling time of the enrichment culture of 11 days, which corresponded to a growth rate μ of $0.062\,d^{-1}$.

4.6.1 Aggregate formation and organisms' distribution

To visualize syntrophic propionate degraders and methanogens in naturally occurring aggregates during propionate oxidation, specific 16S rRNA-based nucleotide probes with Cy3 label (orange fluorescence) for Arc 915-related methanogens and FAM label (green fluorescence) for MPOB1-related pro-

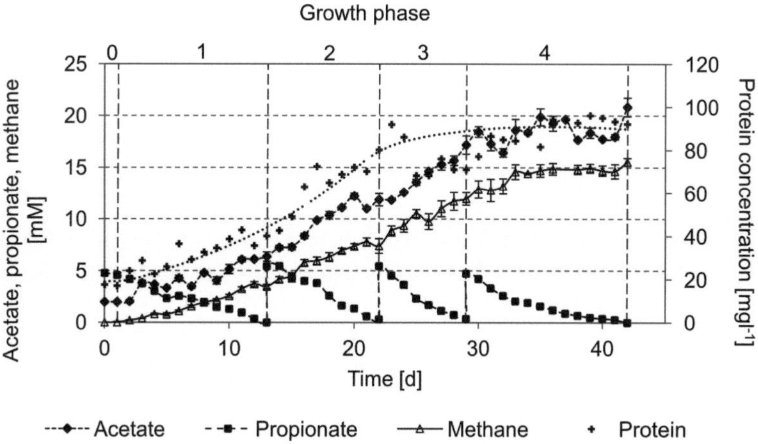

Fig. 4.55: Growth of the mixed culture containing hydrogenotrophic methanogens and *Sytrophobacter*-dominated propionate degraders on propionic acid with acetic acid and CH_4 production; the increase of protein concentration represents cell growth. mM = mmol l^{-1}

pionate oxidizers were hybridized with the respective parent DNA and microscopic images were prepared during different growth phases (Figure 4.56). The total bacterial population was visualized by counterstaining with DAPI (Figure 80, left row). Directly after inoculation in phase 0 and after only 13 days incubation in growth phase 1, single cells could be seen, which aggregated and formed flocs in growth phase 2 to 4. The aggregates were visualized by "3D microscopic images" obtained from a series of 2D images representing different levels of cross section through the aggregates after conversion with the AxioVision software and "extended image definition". The growth of aggregates during the different growth phases can be seen in Figure 4.56 a to j. Initially single cell suspensions were dominant, that were replaced by small aggregates and finally by big aggregates, that were flock-like. Aggregate disintegration in growth phase 3 and 4 is visible in Figure 4.56 h and Figure 4.56 j. The propionate degraders and methanogens were juxtaposed within formed aggregates (inhomogeneous distribution). Micro-colonies composed of propionate-degraders or methanogens were also spotted, but their arrangement in aggregates was not predictable. The spatial configuration of microorganisms corresponded more with the homogeneous distribution model proposed by Schink and Stams (2006) than with the nests- forming model (Figure 1.5).

FISH- and DAPI-images of aggregates taken during different growth phases were analyzed with the use of daime software for digital image analysis in order to calculate the procentual share of propionate degrading and methanogenic microorganisms in the culture. The biovolume fraction of propionate-degraders that hybridized with KOP1- and MPOB1-specific probes, of methanogenic bacteria that hybridized with the Arc 915-specific probe and of prokaryotic organisms that hybridized with the *Bacteria*-specific Eub 338 probe was determined and expressed as percentage of the total population (Figure 81). During the consecutive feeding phases, the proportion of the methanogenic bacteria remained relatively constant, representing approximately 52 % of the total population. The numbers of both examined propionate degrading groups of acetogens were increasing within successive feedings. The KOP1-related biovolume fraction increased from initially 7 % to 29 % after the fourth feeding with propionate, whereas the MPOB1-related biovolume fraction, that included also other propionate degraders (*S. fumaroxidans* and the KOP1-organisms), increased with time from 12 % to 50 % of the total biovolume.

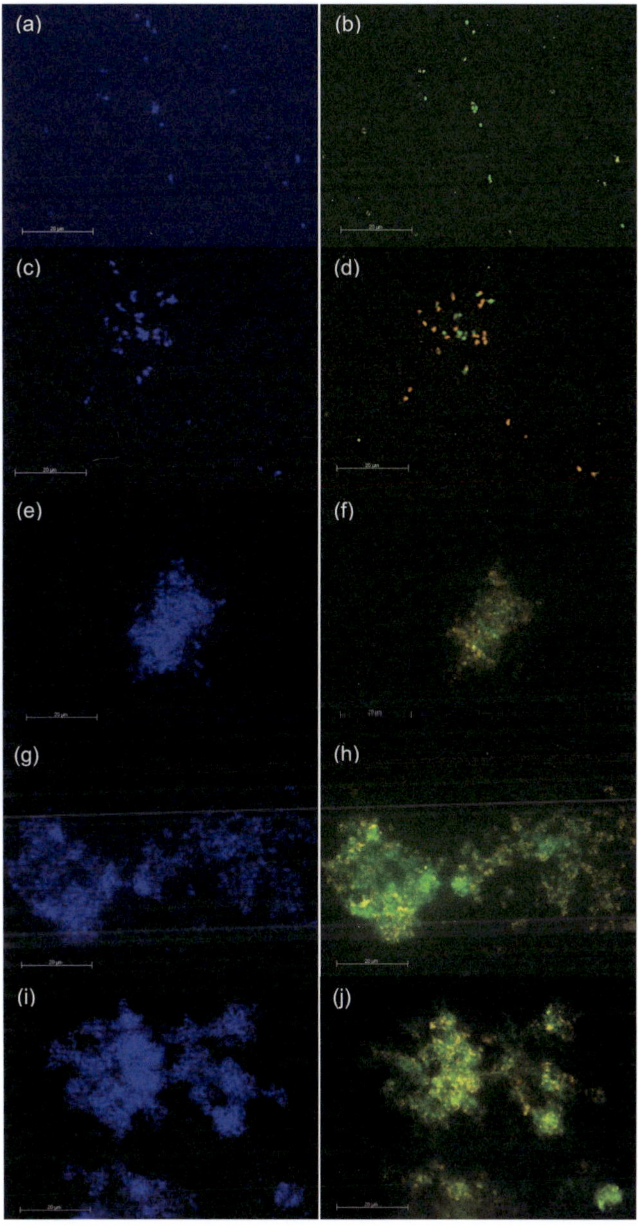

Fig. 4.56: Microscopic images of the propionate-degrading enrichment culture. Left row DAPI-staining of all cells and right row hybridization with a Cy3-labeled Arc915- and a FAM-labeled MPOB1-oligonucleotide probe of the same images showing single cells after inoculation in phase 0 (a ,b), loose floc formation in phase 1 (c, d), aggregate growth in phases 2 and 3 (e-h) and beginning aggregate disintegration in phase 4 (i, j). Bar represents 20 µm.

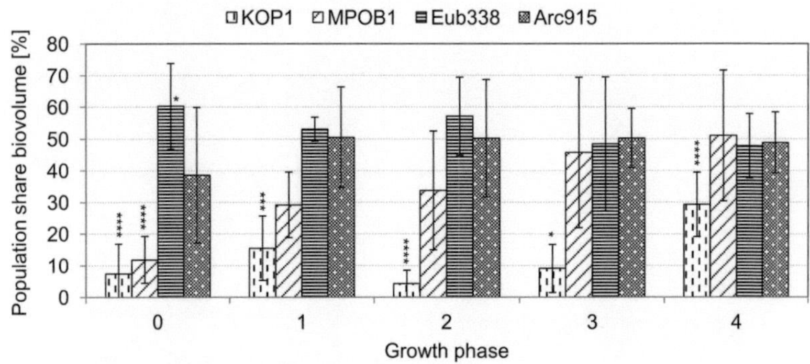

Fig. 4.57: Population analysis of the enriched propionate-degrading culture. The biovolume fractions were estimated as percentage of organisms that hybridized with the mentioned 16S rRNA-targeted oligonucleotide probes compared to the total biovolume determined with DAPI staining. Whereas the probe KOP1 detected *"Syntrophobacter pfennigii"*-related organisms and the probe MPOB1 *"Syntrophobacter fumaroxidans"*-related organisms, the probe MPOB1 gave a signal for both groups. The gene probe Eub338 detected almost all microorganisms belonging to the domain Bacteria and the gene probe Arc915 all microorganisms belonging to the domain Archaea. According to Two-way ANOVA, comparing the mean biovolume fraction determined for a certain organism group with results determined for all other species were $p = 0.03$ for KOP1, $p = 0.007$ for MPOB1, $p = 0.08$ for Eub338, $p = 0.01$ for Arc915; the same test- all measurements vs Eub338 (as a control data set) * $= p < 0.05$, *** $= p < 0.001$, **** $= p < 0.0001$.

Distances measurement between partners

The dimensions of aggregates containing MPOB1-related propionate degraders and Arc 915-related methanogens after successive propionate feedings and the distances between organisms were measured from the "3 D" microscopic images (Figure 4.56). Measurements were performed from outer cell boundary of a propionate degrader to the outer cell boundary of a neighbouring methanogen (positioned at the shortest distance). It can clearly be seen, that interspecies distances between propionate-degraders and methanogens were decreasing with incubation time. At the beginning of the experiment the average distance of single bacteria from each other was 5.3 µm (phase 0). The interspecies distance decreased to 1.56 µm after the first propionate feeding (phase 1), presumably due to aggregation. After the second, third and fourth feeding the interspecies distance was 0.44, 0.53 and 0.29 µm, respectively. Decreasing interspecies distances were reversely correlated with increasing dimensions of the aggregates. Aggregates increased in x, y and z dimensions. The x-parameter (broadness) was taken as a representative size value of the multidimensional aggregate and was plotted against growth (Figure 4.58). The size of the aggregates increased from 18.8 µm to 54.3 µm after 4 feedings. There was almost no increase of the aggregate size during incubation after the fourth feeding, indicating a decreasing stability of the aggregates with a high tendency for disintegration (e.g. Figure 4.56 h, j).

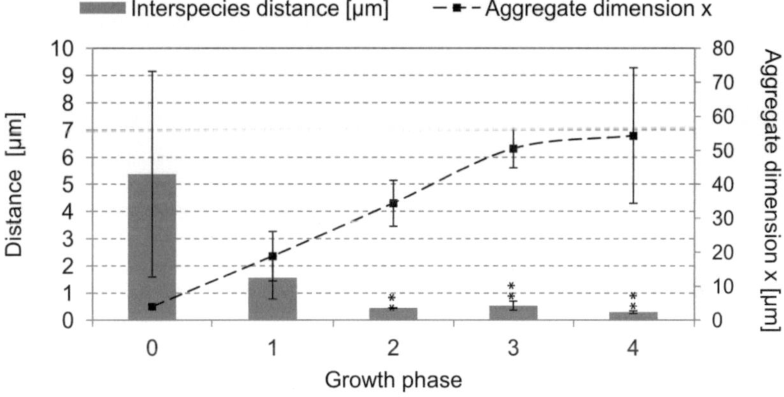

Fig. 4.58: Increasing aggregate dimensions versus decreasing interspecies distances between propionate-degrading and methane producing bacteria during growth phases 0 to 4. ** = $p < 0.01$, one-way ANOVA (distances between organisms in phases 2 to 4 vs mean distance in phase 0); the one-way analysis of variance for the distance measurements within several growth phases gave the value of $p = 0.0009$.

For comparison, during the experimental run I and II the same trend of distances decrease with flock growth increase was observed (Figure 4.59). In run I

(Figure 4.59 a) the distance between propionate degraders and methanogens after inoculation was 2.65 µm and decreased to 0.34 µm in the last phase of experiment. The size of aggregates was simultaneously gaining on dimensions from 0.87 µm in the 2nd phase up to 46.2 µm in the last one. Similarly to previous description, the values were not significantly different for the phase 3 and 4. The smallest interspecies distances measured within the experimental run II (Figure 4.59 b) were 0.26 and 0.29 µm in the last two growth phases (3 and 4), during which the aggregate size increased by 5.8 µm.

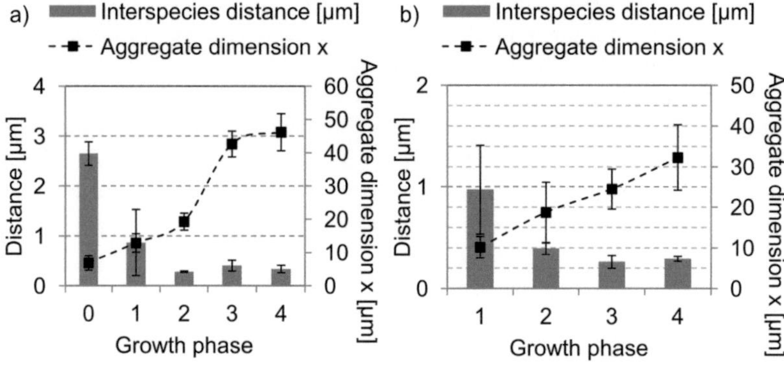

Fig. 4.59: Increasing aggregate dimensions versus decreasing interspecies distances between propionate degrading and methane producing bacteria for a) experimental run I and b) experimental run II; distance measurements in phase 0 were impossible during the run II (b).

With Fick's diffusion law (1.8), the H_2-flux (J_{H_2}) from one organism to another could be calculated. The surface area of the H_2-producing, propionate-degrading organisms (A_{PA}) was calculated from the surface of single organisms (1.58 µm^2) and multiplied by the cell number per ml, the hydrogen diffusion constant in water (D_{H_2}) at 37 °C was 8.27x10^{-5} cm^2/s and the difference of the hydrogen concentration between hydrogen producer (C_{PA} of 9.34 Pa) and hydrogen consumer (C_{H_2} of 0.32 Pa) was set constant with a value of 9.02 Pa. The difference of the hydrogen concentration at the cell surface of the hydrogen producer (C_{PA}) and the hydrogen consumer (C_{H_2}) was calculated according to (4.5) giving the maximum allowed H_2 concentration for propionate oxidation ($\Delta G < 0$) of 9.34 Pa and the minimum H_2 concentration allowed for methane production ($\Delta G > 0$) of 0.32 Pa. The calculated difference between C_{PA} and C_{H_2}, was set as a constant value during the different incubation phases since it was not possible to measure hydrogen in the gas or liquid phase with the necessary low detection limit. The calculated H_2 flux after measuring the interspecies distances was therefore the "maximum possible H_2 flux".

$$\Delta G = \Delta G^\circ + RT\,lnK \qquad (4.5)$$

The Henry constant for solubility of H_2 in water at $37\,^\circ$C was taken as $9.57\,nmol\,l^{-1}\,Pa^{-1}$. The interspecies distance (μm) between the two considered organisms (d_{PAH_2}) was experimentally determined during the respective growth phases. Table 19 shows the calculated maximum possible hydrogen flux J_{H_2} in comparison to propionate oxidation and methane formation rates. The hydrogen flux depended on the hydrogen concentration gradient between producer and consumer and on the interspecies distance. In phase 1, the maximum possible hydrogen flux was $1.1\,nmol/ml/min$. It increased with decreasing interspecies distances to a maximum value in phase 4 of $10.3\,nmol/ml/min$ (Table 4.10). The hydrogen formation rates from propionate oxidation and the hydrogen consumption rates for methane production were considerably lower than the calculated maximum possible hydrogen flux, except for phase 1, were the difference between the calculated maximum possible hydrogen flux and the measured hydrogen production/ consumption rate were approximately in the same range (Table 4.10).

Table 4.10: Comparison of hydrogen production rates, hydrogen consumption rates and the hydrogen flux from the propionate-oxidizers to the methanogens in the syntrophic propionate degrading culture.

Growth phase (culture age [d])	Hydrogen formation rate[a] [nmol/ml/min]	Hydrogen consumption rate[a] [nmol/ml/min]	Hydrogen flux $J_{H_2}^{c)}$ [nmol/ml/min]
1 (13)	0.746 ± 0.070	0.741 ± 0.030	1.10
2 (22)	1.375 ± 0.021	1.486 ± 0.280	4.48
3 (29)	1.572 ± 0.017	1.583 ± 0.237	5.06
4 (42)	1.239 ± 0.112	1.111 ± 0.208	10.30

[a] The hydrogen formation rate was calculated from the propionate oxidation rate assuming, that per mol of propionate 3 mol of H_2 were produced (data Figure 4.60 a).

[b] The hydrogen consumption rate was calculated from the methane formation rate assuming, that per mol of methane 4 mol of H_2 were consumed (data from Figure 4.60a).

$^{c)}$ J_{H_2} calculated according to Fick's diffusion law, cell numbers of MPOB1-related propionate degraders in phase 1 were 5.05 x10^7 ml^{-1}, in phase 2 5.83 x10^7 ml^{-1}, in phase 3 7.91 x10^7 ml^{-1} and in phase 4 8.82 x10^7 ml^{-1}, cell numbers were calculated from the total cell number and percentage of MPOB1-related propionate degraders in each growth phase from Figure 4.57.

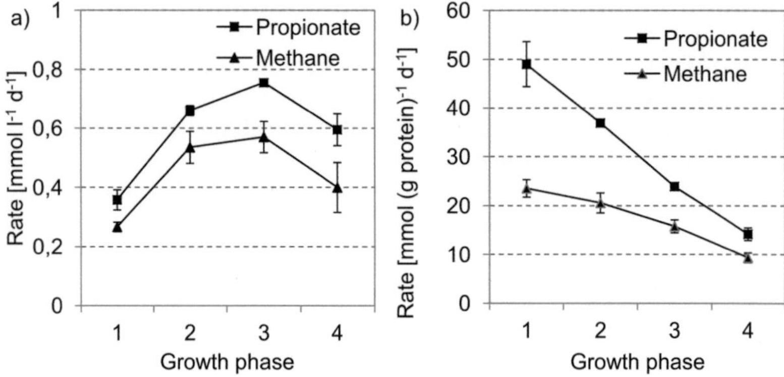

Fig. 4.60: The rates of a) propionate oxidation and methane formation with b) specific propionate oxidation and methane formation rates within 4 feeding phases, where specific rates were obtained from the respective total protein concentration of all bacteria from Figure 4.55 and calculating the protein content of the propionate degraders and of the methanogens according to their biovolume fraction in phases 1 – 4 after MPOB1- specific and Arc 915- specific labeling (Figure 4.57).

4.6.2 Acetate as an inhibiting agent during propionate degradation

Anaerobic propionate oxidation was coupled with hydrogen formation, but proceeded only at low hydrogen partial pressure and thus was dependent on hydrogenotrophic methanogens. The propionate-degradation rates should match with methane production rates, as deductable from (1.1) and (1.3) (0.75 mol methane produced from 1 mol propionate oxidized). Figure 4.60 a shows the propionate oxidation and methane formation rate from H$_2$/ CO$_2$ during the different growth phases. Both rates were highest in phase 3 with maximum values of 0.76 and 0.57 mmol l^{-1} d^{-1}, respectively. The specific propionate-degradation and methane formation rates are shown in Figure 4.60

b. To obtain the specific rates, the total protein concentration was attributed to propionate degraders and methanogens by hybridizing the mixed culture with 16S rRNA-based oligonucloetide probes targeting MPOB1-related propionate degraders and Arc 915-related methanogens. As a reference an unspecific staining of all organisms with DAPI was performed and the results were expressed as "biovolume fraction" (Figure 4.57). As can be seen in Figure 4.60 b, the specific propionate oxidation and methane formation rates decreased linearly during increasing time of incubation. The maximum specific propionate oxidation and methane formation rate was 49 and 23 $mmol\,mg^{-1}\,d^{-1}$, respectively. These decreasing specific propionate oxidation and methane formation rates may indicate an inhibition of syntrophic propionate degradation in the enrichment culture.

To test whether accumulating acetate inhibited propionate oxidation, increasing concentrations of acetate (5, 10, 15 and $20\,mmol\,l^{-1}$) were added and the propionate degradation rate with an initial propionate concentration of $5\,mmol\,l^{-1}$ was measured (Figure 4.61). The slopes obtained for assays with increasing acetate concentration (Figure 4.61 a) gave the decreasing propionic acid degradation rates (Figure 4.61 b). The experiment was simulating the conditions during the successive growth phases in Figure 4.55. Without external acetate addition, $0.32\,mmol\,l^{-1}\,d^{-1}$ propionate was oxidized, which corresponded well with the results shown in Figure 4.60 a, where the propionate oxidation rate in growth phase 1 was $0.35\,mmol\,l^{-1}\,d^{-1}$ $(= 100\,\%)$. With the addition of $5\,mmol\,l^{-1}$ acetate the propionate oxidation rate decreased to $77.7\,\%$ and in the presence of 10, 15 and $20\,mmol\,l^{-1}$ acetate to 60.7, 57.9 and $52.9\,\%$ respectively (Figure 85 b), indicating end product inhibition.

4.6.3 Formate versus hydrogen interspecies transfer

Hydrogen and/ or formate may serve for interspecies electron flow between syntrophic propionate degraders and methanogens. Since both partners might express formate dehydrogenase, formate might be produced from $NADH + H^+$ at a concentration of less than $0.01\,mmol\,l^{-1}$ and a redox potential of $< -290\,mV$ (Sieber et al. 2012) or from H_2 and CO_2. Vice versa it might be degraded by both organisms. To test, whether formate was released by the propionate oxidizers and consumed by the methanogens in the syntrophic enrichment culture, batch tests with $H_2:CO_2$, formate and propionate in the absence and presence of the methanogenic inhibitor bromoethanesulfonic acid (BESA) were performed. Methane formation (Figure 4.62 a) and intermediarily released formate (Figure 4.62 b) were analyzed. If H_2/

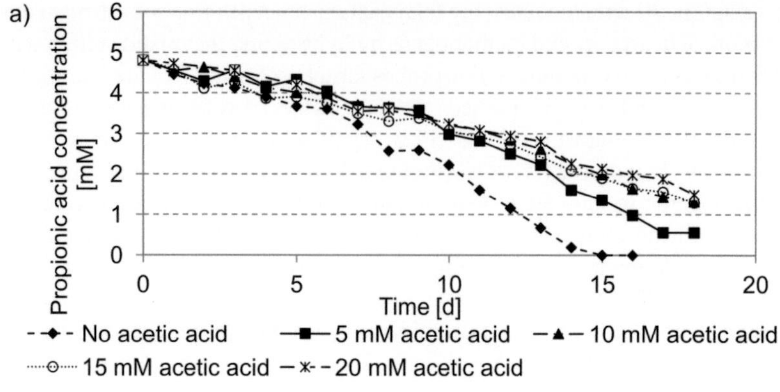

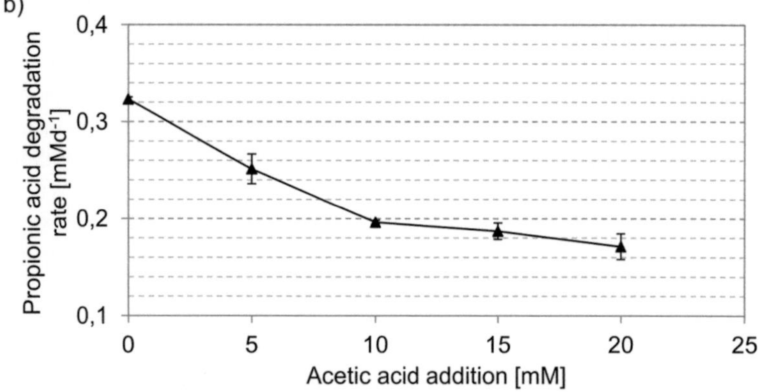

Fig. 4.61: Inhibiting effect of acetic acid on propionate degradation; the initial propionate concentration was $5\,\text{mmol}\,l^{-1}$ in each assay; different acetate concentrations were added to each assay; a) propionic acid degradation slope behavior in relation to different acetic acid concentration and b) resulting degradation rates. $mM = \text{mmol}\,l^{-1}$

CO_2 was supplied to the mixed culture methane and significant amounts of formate were the products. Whereas methane must have been generated by the methanogens, formate might have been generated by both, methanogens and propionate degraders. In BESA-inhibited assays no methane but some formate was produced in the presence of H_2/CO_2 and propionate. Externally added formate was cleaved to methane and CO_2 by the methanogens with a rate of $0.42\,\text{mmol}\,l^{-1}\,d^{-1}$, which was 6 times lower than the methane production rate from $H_2:CO_2$ ($0.26\,\text{mmol}\,l^{-1}\,d^{-1}$). In the assay with propionate the methane formation rate from degradation products H_2 and CO_2 (aceticlastic methanogens were not present in the mixed culture) was $0.037\,\text{mmol}\,l^{-1}\,d^{-1}$, due to slow propionate degradation. No formate was detectable as a product or intermediate. Some formate was, however, produced from propionate when methanogenesis was inhibited completely by BESA (Figure 4.62 a, b). Whether the propionate degraders generated the formate already or whether they excreted H_2/CO_2 and the formate was formed by the methanogens via formate dehydrogenase could not be distinguished. However, it is probable that formate was generated exclusively or predominantly by the methanogens, since they were able to form a similar amount of formate from H_2/CO_2 during complete inhibition of methanogenesis with BESA (Figure 4.62 a, b). A direct proof of interspecies hydrogen transfer during propionate oxidation with and without BESA failed due to the detection limit of hydrogen with the used GC-TCD of 128 Pa and presumably its conversion to formate at less than 128 Pa hydrogen partial pressure, when methanogenesis was inhibited by BESA. No methane formation was observed after addition of BESA to all assays indicating the successful inhibition of the methanogens (Figure 4.62 a).

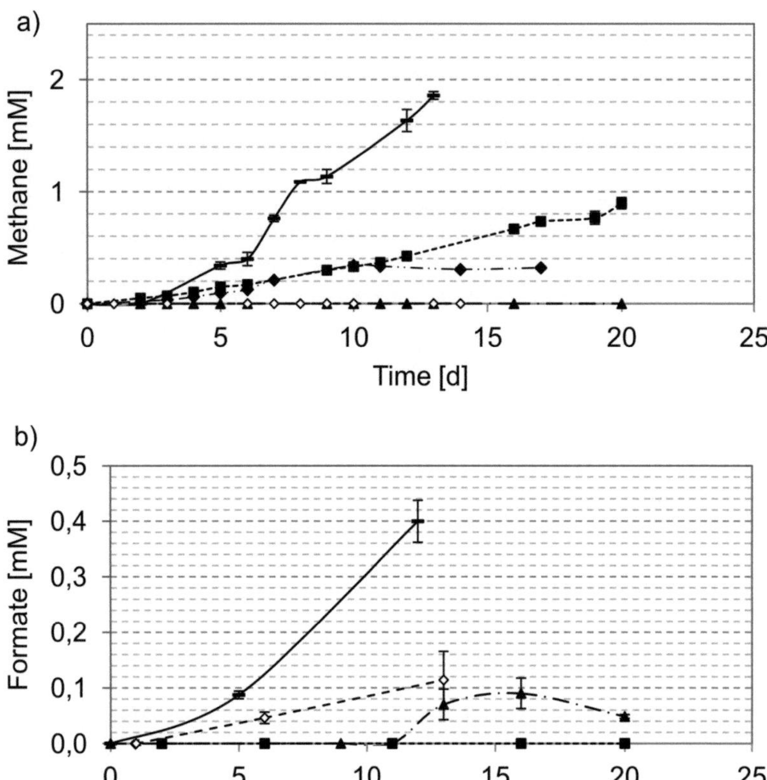

Fig. 4.62: Production of a) methane and b) formate by the syntrophic culture after addition of respective substrates ± BESA as an inhibiting agent for methanogens. mM = mmol l^{-1}

5 Discussion

5.1 Analytical analysis of volatile fatty acids

5.1.1 Sample preparation

Sample acidification prior analysis is a common approach during preparation (Namieśnik and Jamrógiewicz 1998). For the analysis of organic compounds, acids like H_2SO_4 (EPA, APHA, ASTM recommendation) or H_3PO_4 (ISO recommendation) are used for sample conservation by pH lowering to values below 4 or 2. This should prevent the precipitation of metal oxides and hydroxides as well as inhibit the biological activity. Sample acidification decreases also the VFAs solubility in water. If the pKa of an acid is comparable with the pH of a solution, the concentration of dissociated and undissociated form of this compound are equal. Lowering the pH by 2 units in comparison to pKa reduces the dissociated form of a compound content to 1 % (Banel 2010). Besides, salt addition was also reported to have the lowering effect on the VFAs solubility in water (Pan et al. 1995, Banel et al. 2011). According to the above, lowering the pH of a sample and salt addition enhances the undissociated acid compounds transfer to the head phase. During this study the application of hydrophosphoric acid and $NaHSO_4$ solution was compared. The use of the second agent combined both, sample acidification and salt addition effect and resulted in higher recovery of analytes. Obtained peak area for acetic acid and propionic acid were, respectively, 33 % and 22 % higher with $NaHSO_4$ than with H_3PO_4 addition (Figure 4.4).

Additionally, when the sample was acidified before centrifugation and biomass pellet rejection, the recovery of a sample was higher. The difference between peak areas measured was 11 % for acetic acid and 9 % for propionic acid (Figure 4.2). Lowering the pH of a liquid containing the biomass forced the diffusion of acids through the cell wall. This was crucial for mass balance calculation.

5.1.2 Headspace analysis versus direct sample injection

Direct injection of wastewater samples on column for gas chromatographic analyses is not recommended; because such liquid samples apart from VFAs contain usually other organic substances can influence the result of analysis. Influencing factors include detrimental effects like background noise elevation in the system, analytical column dysfunction or system contamination (Peldszus 2007). The resolution of resulting chromatograms would decline and hence the detection limits for analytes could be elevated. Moreover, water introduced to the system shortens the life of the stationary phase, especially polyethylene glycol that is recommended for VFAs separation and was used during this study. Background noise elevation can be expressed by the raise of a baseline, which was observed during the method development (Figure 4.14). That hindered the resulting chromatogram resolution and VFAs detection. High water content present in the system additionally affected the resulting ionization fragments of analytes that were different from the NIST library spectra (Figure 3.9). These differerences between main fragmentation ions [m/ z] were as follows: acetic acid: 43, 45, 60 (NIST) and 43, 46, 61 (direct injection); propionic acid: 45, 57, 74 (NIST) and 46, 59, 75 (direct injection).

Headspace analysis is considered as the sample preparation method that is characterized by high detection limits of analyzed compounds (Cruwys et al. 2002; Gonzales-Fernandez and Garcia-Encina 2009). Nevertheless, the advantages of the method like simple separation of analytes from gaseous phase (after primary solid- or liquid-matrix change) and the possibility of direct carrier gas/ analyte vapour mixture injection into GC (Namieśnik and Jamrógiewicz 1998) make it still attractive for certain applications.

The detection limits
The most popular approach for VFAs analysis found in the literature is the GC-FID combination, where the limits of detection for acetic acid and propionic acid are strongly dependant on the way of sample preparation before the analysis. Cruwys et al. (2002) reports the detection limit for the HS-GC-FID system with FFAP analytical column for acetic- and propionic acid to be $0.06\,\mathrm{mmol\,l^{-1}}$ and $0.04\,\mathrm{mmol\,l^{-1}}$, respectively, with no sample derivatization. Samples derivatization with methanol and sulfuric acid addition gave a detection limit of $5\,\mathrm{mg\,kg^{-1}}$ for acetic acid and $1\,\mathrm{mg\,kg^{-1}}$ for propionic acid in HS-GC-FID analysis of compost samples with Rtx-1701 analytical column (Himanen et al. 2006). The MS application as a detector was also reported for VFAs analysis application in several combinations. The most popular

combination seems to be the HS-SPME-GC-MS with different ionization types: positive chemical ionization (PCI), negative chemical ionization (NCI) or electron ionization (EI) (Abalos et al. 2000; Larreta et al. 2006). With the SPME the detection limits for acetic acid and propionic acid were, respectively, 2.5 μmoll^{-1} and 0.07 μmoll^{-1} (NCI), 1.9 μmoll^{-1} and 0.3 μmoll^{-1} (PCI), according to Abalos et al. (2000). Larreta et al. (2006) reported the detection limits of 0.02 mmoll^{-1} for both acids with EI. The intention of this study was to meet the requirements for differentiation of ^{12}C and ^{13}C isotopes containing fragments of acetate or propionate, so the MS detector had to be applied. The detection limits for the HS-GC-MS (EI) method developed during this study were 0.09 mmoll^{-1} for acetic acid and 0.02 mmoll^{-1} for propionic acid. Considering the fact, that the samples were not derivatised or prepared by SPME for the analysis, the result is satisfactory and allows anaerobic pathway studies.

In comparison with the direct injection, where the HS preparation step of the sample was excluded and the GC-MS combination was applied, the LOD value for acetic acid was reduced ten times to 0.008 mmoll^{-1} and in the case of propionic acid down to 0.011 mmoll^{-1} (Table 4.5). The mass spectra of analytes were of a good resolution, giving the expected fragmentation ions (Figure 4.13). Changing the sample matrix from methanol (used for standard solution preparation) to water influenced the ionization, what resulted in shifting the main fragmentation ions. That precluded the method as suitable for metabolic pathway determination. Furthermore, the significantly elevated water vapour content in the system affected the baseline position in produced chromatograms (comparing Figure 4.12 and Figure 4.14) and, consequently, the sensitivity of the method. The liquid-liquid extraction of samples was not considered due to poor analytes' recovery, large solvent volume demand, production of relatively large waste amount and time consumption.

Calibration curve and coefficient of variation
The obtained calibration curves (Figure 4.11) for acetic acid and propionic acid and their regression coefficient are in a good correlation with the values obtained during similar studies of Banel (2010). Comparison between HS-GC-MS method developed during this research and other methods, like GC-FID, HS-GC-FID or HS-SPME-GC-MS with polydimethylosiloxane-carboxen (PDMS/ CAR) fiber coating is presented in Table 5.1. The coefficient of variation (CV) was calculated from Equation (5.1).

$$CV = \frac{S \cdot 100\%}{\overline{x}} \tag{5.1}$$

113

S - standard deviation

$\bar{x}$ - arithmetic mean

Table 5.1: Characteristics of chosen methods for VFAs analysis.

Method	Parameter	Acetic acid	Propionic acid	Reference
HS-GC-MS	Concentration range	$30\,\mathrm{mg\,l^{-1}}$ to $300\,\mathrm{mg\,l^{-1}}$	$37\,\mathrm{mg\,l^{-1}}$ to $370\,\mathrm{mg\,l^{-1}}$	This study
		$0.5\,\mathrm{mmol\,l^{-1}}$ to $5.0\,\mathrm{mmol\,l^{-1}}$		
	R^2	0.9959	0.9993	
	Average CV [%][1]	6.8	7.0	
GC-FID	Concentration range	$50.0\,\mathrm{mg\,l^{-1}}$ to $500\,\mathrm{mg\,l^{-1}}$		
		$0.8\,\mathrm{mmol\,l^{-1}}$ to $8.3\,\mathrm{mmol\,l^{-1}}$	$0.7\,\mathrm{mmol\,l^{-1}}$ to $6.8\,\mathrm{mmol\,l^{-1}}$	
	R^2	0.9920	0.9910	Banel 2010
	CV [%]	0.76 to 5.52	0.54 to 4.86	
HS-GC-FID	Concentration range	$25\,\mathrm{mg\,l^{-1}}$ to $100\,\mathrm{mg\,l^{-1}}$	$5\,\mathrm{mg\,l^{-1}}$ to $300\,\mathrm{mg\,l^{-1}}$	
		$0.4\,\mathrm{mmol\,l^{-1}}$ to $1.7\,\mathrm{mmol\,l^{-1}}$	$0.07\,\mathrm{mmol\,l^{-1}}$ to $4.1\,\mathrm{mmol\,l^{-1}}$	
	R^2	0.9933	0.9998	
	CV [%]	0.47 to 15.0	1.66 to 16.3	
HS-SPME-GC-MS (PDMS/CAR)[2]	Concentration range	$10.0\,\mathrm{mg\,l^{-1}}$ to $200\,\mathrm{mg\,l^{-1}}$	$12.0\,\mathrm{mg\,l^{-1}}$ to $154\,\mathrm{mg\,l^{-1}}$	
		$0.17\,\mathrm{mmol\,l^{-1}}$ to $3.3\,\mathrm{mmol\,l^{-1}}$	$0.16\,\mathrm{mmol\,l^{-1}}$ to $2.1\,\mathrm{mmol\,l^{-1}}$	
	R^2	0.9996	0.9930	
	CV [%]	0.70 to 4.42	1.07 to 3.40	

[1] Calculated out of 25 experimental runs

[2] PDMS/ CAR = polydimethylosiloxane-carboxen fiber coating

According to the above comparison (Table 5.1), the developed method is characterized with a good reproducibility. The HS application for sample preparation seems to be the reason for diminished precision of measurements, when the GC-FID and HS-GC-FID are taken into consideration. The difference between highest values of variation coefficient between these two equals about 10 % for acetic acid and 11 % for propionic acid. Solid phase microextraction combined with HS appears to be a solution for measurement precision improvement. However, according to quite high metabolites' concentration in the experiments performed during this study (in some cases above 20 mmol l^{-1}), this was not necessary.

5.1.3 HS-GC-MS as optimal method choice

With the use of the developed HS-GC-MS, the time course of VFA concentrations during ^{13}C labelled propionic acid degradation can be measured without time-consuming sample purification and/ or derivatisation within a short analysis time. Furthermore, the method may be used to elucidate metabolic pathways, when VFAs are involved. This allows at least restricted conclusions as to the microbial communities and/ or species involved in anaerobic digestion and facilitates their identification just like e.g. observing organisms' morphology.

5.2 Metabolic pathway identification and mass balance for propionic acid degradation experiments

The operational parameters for the whole HS-GC-MS system were successfully optimized to obtain mass spectra of a good resolution (Figure 4.7), what granted confident identification of ^{13}C-labeled compounds, and hence, confident anaerobic pathway identification. To secure this statement, ^{13}C-labeled metabolites (Table 4.2) were analyzed, detected and successfully identified (Figures 4.9 and 4.10).

Method application for the microorganisms collected from different habitats resulted in their description with the propionic oxidation pathway identification. Mass spectra of acetic acid obtained during 1-^{13}C- labeled propionic acid oxidation by each examined culture did not point to ^{13}C isotope incorporation in the produced compound moiety (methyl- and/ or carboxyl- group). That indicates the methyl-malonyl-CoA pathway as dominant in each microbial consortium:

- originating from industrial-scale reactor treating biowaste, fresh sludge (sample A);

- originating from industrial-scale anaerobic reactor treating biowaste, enrichment culture (sample B);

- originating from lab-scale anaerobic reactor LIPI treating market waste, directly collected sample (sample C); (No. 6, Table 4.6);

- originating from lab-scale anaerobic reactor LIPI treating market waste; enrichment culture (sample D);

- originating from lab-scale anaerobic reactor ITB treating market waste; directly collected sample (sample E); (No. 3, Table 4.6).

According to Figure 3.7, the compound "labeling" should be noticeable in the CO_2 produced by propionic acid degraders. Unfortunately, the determination of carbon dioxide was not possible. To secure the result obtained by MS metabolites' detection, the mass balance for the experiments according to methyl-malonyl-CoA metabolic pathway of propionate degradation has to be considered. It states that if 1 mol propionic acid is oxidized, 1 mol acetic acid and 0.75 mol methane ((1.1) and (1.3)) will be produced. It is also possible that resulting acetate will be converted into methane according to (1.2) e.g. by *Methanosaeta sp.* or *Methanosarcina sp.*. The concentration values of substrate added and metabolites produced are summarized in Table 5.2.

Table 5.2: Mass balance for ^{13}C-labeled propionic acid degradation experiments by different microbial consortia.

Sample	Propionic acid	Acetic acid	Methane	Incubation time [h]	Surplus methane[3]
A	9.6±0.6 mmol l^{-1}	9.5±0.6 mmol l^{-1}	0.45±0.02 mmol l^{-1}[1]	453	-
B	6.68±0.09 mmol l^{-1}	6.63±031 mmol l^{-1}	5.52±0.10 mmol l^{-1}	102	2%
C	5.27±0.07 mmol l^{-1}	4.93±0.30 mmol l^{-1}	4.13±0.14 mmol l^{-1}	50	4%
D	86.76 µmol[2]	63.77±4.46 µmol	90.60±2.71 µmol	194	3%
E	24.67 µmol[2]	17.55±0.21 µmol	28.20±0.85 µmol	94	9%

[1] Methane production of a culture with inhibited methanogenesis (BESA addition)

[2] Substrate initially measured concentration minus the remaining substrate measured at the end of experiment (D: after 194 h; E: after 94 h)

[3] Maximum acceptable deviation for a mixed culture under experimental conditions should not exceed 10%

In the fresh sludge experiment (A) the stoichiometry of substrate to product, after considering the measurement error is 1:1, which is in good correlation to the assumption that from 1 mol propionate, 1 mol acetate is produced. Methane production was inhibited with BESA in order to inhibit acetate degradation by aceticlastic methanogens, so only a marginal CH_4 production was recorded. Organisms from enrichment culture B (enriched out of A) produced acetic acid from propionate at a ratio 1:1, and additionally the methane production was $0.46\,mmol\,l^{-1}$ higher than theoretically expected, what equals 2 % deviation and is a generally acceptable error for a mixed culture under experimental conditions. That supports the statement, that the main propionic acid degradation pathway for the examined culture is the methyl-malonyl-CoA pathway, although some cells of C-6-dismutating *S. propionica* might have been present in the sample A (Figure 4.18).

Samples originating from LIPI contained aceticlastic methanongens (Figure 4.29), and consequently the ratio of produced acetate to fed propionate was not 1:1. In the mixed culture C, the "lacking" acetic acid must have been converted into methane, and so the theoretically expected CH_4 production is $4.29\,mmol\,l^{-1}$. That gives 4 % deviation from the measured concentration and is within acceptable error. The balance for the sample D (culture enriched previously on propionate) had to be done similarly, where the acetic acid amount converted into methane was $22.99\,\mu mol$ ($86.76\,\mu mol$ propionate minus $63.77\,\mu mol$ produced acetate). Theoretically there should be $88.06\,\mu mol$ methane produced ($65.07\,\mu mol$ by hydrogenotrophic methanogens and $22.99\,\mu mol$ by aceticlastic methanogens). The deviation of 3 % was stated in relation to measured value (Table 5.2) and is within acceptable error. The mass balance confirms the methyl-malonyl-CoA pathway as dominant for examined organisms.

ITB originating culture (E) contained aceticlastic methanogens (Figure 4.27), and hence, the ratio between produced acetate from fed propionate was not 1:1. Missing acetate of $7.12\,\mu mol$ was transferred into methane, giving the theoretical methane production amount of $25.62\,\mu mol$ (Table 4.7). Measured methane surplus equals $2.58\,\mu mol$, what results with 9 % deviation. The methyl-maolonyl-CoA pathway was domnianant during propionic acid oxidation by for this culture. The enrichment culture on propionate could not be analyzed with HS-GC-MS method for metabolic pathway determination because of a very long substrate degradation time and almost no acetic acid production (Figure 4.27). One of the possibilities to conduct the experiment was addition of BESA as an inhibiting agent to the sample, which was undertaken. This approach unfortunately extended the degradation time even more (alike the difference between incubation times of the samples A and B, Table

5.2). As there had been no propionic acid degradation recorded after 30 days of sample incubation, the experiment was terminated.

The obtained data give unequivocal result, that in all examined cultures, the directly collected and enriched ones, the dominant digestion pathway of propionic acid was the methyl-malonyl-CoA pathway described by Plugge et al. (1993). Until now the only organism known to degrade propionate according to the alternative C-6-dismutation pathway (Figure 3.7) with butyrate intermediate production is *Smithella propionica* (de Bok et al. 2001). To determine whether the collected samples included the *S. propionica* organisms the 16S rRNA- based Smi sr 354 oligonucleotide probe was used during FISH (Ariesyady et al. 2007). The probe gave positive signals for "fresh sludge" culture (Figure 4.18), however, as it is specific for *S. propionica* and members of the *Syntrophus* group, it is not explicit, that species producing labeled acetic acid from 1-^{13}C -propionate were present in the examined consortium. If that was the case, the content of *S. propionica* cells must have been too low to be detected with the HS-GC-MS method for pathway determination. That would be consistent with the results described by Narihiro et al. (2012), where from 6 different anaerobic digesters (2 mesophilic full-scale upflow anaerobic sludge blankets UASB for wastewater treatment, a thermophilic pilot-scale UASB reactor for industrial wastewater treatment; mesophilic lab-scale UASB and an expanded granular sludge blanket (EGSB) treating industrial wastewater and municipal sewage, as well as mesophilic full-scale anaerobic digester treating municipal sewage), only one contained *S. propionica* and its content according to results of sequence specific cleavage of small subunit (SSU) rRNA with RNase H was reaching nearly 4 %. Most of the syntrophic bacteria detected by the method were from the *Syntrophobacter* genus (found in two reactors at a level of 3 to 4 %) and *Pelotomaculum* genus (found in a thermophilic reactor at a 3.5 % content). Studies on full-scale biowaste fermenter population shifts (within a year of performance) done by Malin and Illmer (2008) do not report presence of any propionic acid degraders using the DGGE as a detection tool (species present at a content of 1 % of a total population can be detected by this method). Out of five examined digesters by Oude Elferink et al. (1998), *Syntrophobacter sp.* and *Desulforhabdus amnigenus* were found in three of them: a full-scale UASB treating papermill wastewater at a level of 0.3 % to 0.7 %, and two lab-scale ones fed with sucrose and VFAs at 1.5 % to 2.5 % of total bacterial population in both. That implies generally quite diminutive syntrophs presence in digesters.

Microbial analysis of inocula from a sewage sludge digester used for the start-up of differently fed mesophilic laboratory scale continuously- and minimally mixed reactors showed a *Smithella propionica* content of 0.3 % or below the detection limit (0.2 %). *Syntrophobacter* propionate oxidizers were present at 1 % to 1.4 % level (% SSU rRNA). These values changed to respectively 0.6 % and $\approx$ 1 % during start up, $\approx$ 0.2 % and $\approx$ 1.2 % to 1.5 % at steady state and up to 2 % and $\approx$ 1.6 % to 1.9 % (depending on the organic loading rate) during organic overload. These results presented by McMahon et al. (2004) also indicate relative small percentage of syntrophs capable of propionate degradation, especially *S. propionica*, taking part in anaerobic degradation of municipal solid waste and sewage sludge.

5.3 Microbial isolation

Isolation of propionic acid degraders in a pure culture is very difficult and time consuming because of long doubling times of organisms (Table 5.4), the fact that they are obligate anaerobes and in some of the cases obligate syntrophs (e.g. *Pelotomaculum schinkii*, (de Bok et al. 2005)). Apart from the fact, that probably 1 % of the microflora is cultivable at all, most syntrophic acetogenic bacteria are not able to grow in agar shake cultures or on plates. Stams et al. (1993) e.g. could not isolate organisms growing on propionate plus fumarate in agar shake cultures, as they did not form colonies.

The most successful enrichments during this study were obtained after sulfate addition to cultures grown on propionate (Table 4.8). Sulfate addition as an enhancing agent for propionate oxidation was confirmed by e.g. Briones et al. (2009), who determined favored propionate degradation after sulfate addition to lab-scale mesophilic digesters inoculated with organisms from an industrial reactor of a brewery. This phenomenon was explainable by the fact, that almost all known propionate oxidizing bacteria were sulfate reducers. Furthermore, almost all described propionate degraders are sulfate reducers (Table 1.2) and propionate oxidation coupled with SO_4^{2-} reduction seems to have more advantages resulting in a lower $\Delta G^{\circ\prime}$.

The propionic acid degrading organisms obtained by enrichment on propionate were identified as *Syntrophobacter* organisms capable of sulfate reduction. The *Syntrophobacter* group was found to be the most abundant one among propionate degraders present in anaerobic digesters (as discussed in previous subsection 5.2) and the species described in the literature are *S. fumaroxidans* (Harmsen et al. 1998), *S. wolinii* (Liu et al. 1999), *S. pfennigii* (Wallrabenstein et al. 1995) and *S. sulfatireducens* (Chen et al. 2005). Zellner et al. (1996) described also a strain HP1.1 of the genus *Syntrophobacter*.

Molybdate inhibition of propionate oxidation in the mesophilic tri-culture

(with no methanogens) was a confirmation of parallel SO_4^{2-} reduction requirement for the process to occur. However it did not clear whether the propionate
oxidizer itself was responsible for sulfate reduction, or the vibrios present
in the ecosystem are partner organism for syntrophic growth. Zellner and
Neudörfer (1995) described propionate degradation by a biofilm consisting
of mainly *S. wolinii* morphologically distinct organisms and methanogens
in anaerobic fluidized bed reactors, where molybdate addition completely
inhibited the process. That implied propionate degraders to be sulfate reducers. On the other hand, sulfate addition to a non-inhibited culture did
not enhance propionic acid oxidation, and 78 % electron flow was still occuring independently of sulfate presence (via interspecies hydrogen transfer
with CH_4 production), while 22 % went into sulfate reduction. Nevertheles,
the fact that organisms from a culture enriched during this study hybridized
with SRB oligonucleotide probe was the final confirmation that the lemon-
shaped propionate degrader, which also gave a positive signal after Synbac
824 probe hybridization, was capable of sulfate reduction, like all known
members of genus *Syntrophobacter* (Table 1.2). Mesophilic *Syntrophobacter*
hybridized also with MPOB1 (*S. fumaroxidans* and *S. pfennigii* specific), but
no KOP1 (*S. pfennigii* specific probe), what might indicate that this bacteria
could represent close relatives of *S. fumaroxidans*. *S. pfennigii* is capable
of gas vacuole formation (Wallrabenstein et al. 1995) just like the member
culture enriched during this study (Figure 4.36), what on the other hand would
support the bacterium relatedness to KOP1. The size of the lemon-shaped
microbe (Table 4.8) was slightly different than that of *S. pfennigii* being 1.0
to 1.2 x 2.2 to 3.0 µm (Wallrabenstein et al. 1995). Nevertheless, it was in
good correlation within cell sizes reported for other *Syntrophobacter sp.*, like
e.g. *S. sulfatireducens* (1.0 to 1.3 x 1.8 to 2.2 µm, Chen et al. 2005) or *S.
fumaroxidans* (1.1 to 1.6 x 1.8 to 2.5 µm, Harmsen et al. 1998).

The thermophilic enrichment culture obtained after pasteurization of the bacterial suspension and consisted of propionate degrading bacteria identified as
a *Syntrophobacter sp.* (hybridized with Synbac 824 probe) and a methanogenic partner, which was not capable of degrading any substrate other than $H_2/
CO_2$. The culture degraded propionic acid with and without sulfate addition
and, not alike other propionate degraders, oxidized the acid at the same rate in
both cases (Table 5.3). With inhibited methanogenesis, the growth on propionate and sulfate was very slow (Figure 4.43 c), and with molybdate addition
the bacteria were completely inhibited. This suggests the propionate degrader
to be a sulfate reducer. Both known thermophilic *Desulfotomaculum sp.* are
also sulfate reducers, however their localization in the phylogenetic tree of
syntrophic bacterial species is distant from *Syntrophobacter sp.* (Sieber et al.

2012). It is possible, that the member of the thermophlic culture is a new thermophilic species within the *Syntrophobacter* group. Works on isolation of the methanogenic partner, whose optimal growth temperatue was 65 °C, were undertaken. However, the DNA sequence for its classification was not yet determined at the end of this study.

Sulfate reduction and propionate degradation under halophilic conditions were noticed by the enrichment culture on propionate and sulfate from the North Sea. The degradation rate for this substrate was, however, very low, but this result is comparable with previous research done by other scientists. There are a few fermentative halophilic bacteria known until now. Basso et al. (2009) analyzed organisms collected from saline, slightly alkaline environment- a deep subsurface aquifer by determining, inter alia, the MPN counts in anaerobic media incubated at 37 °C, where 2.5 sulfate reducing bacteria growing on propionate per ml were recorded. Methanogenic bacteria growing on H_2/CO_2 (95 bacteria ml^{-1}) or acetate/ formate (0.9 bacteria ml^{-1}) were also present. The SRB activity was reported in the upper layer of marine Arctic sediments (reduction rate 0.14 fmol per cell per day) (Ravenschlag et al. 2000). On the other hand, cultivation of sulfate reducing bacteria and methanogens originating from deep-sea anoxic basins by Brusa et al. (1997) failed. Desulfobaba fastidiosa was isolated from the sulfate-methane transition zone of marine sediments and described as a SRB propionate degrader with a growth rate $\mu = 3.85d^{-1}$ by Abildgraad et al. (2004). In comparison with other propionate degraders (Table 5.3), its growth was extremely fast. *Desufobaba gelida* (Knoblauch et al. 1999) and *Desulfomusa hansenii* (Finster et al. 2001) were also halophilic representatives capable of propionate degradation to acetate and carbon dioxide. The first one has a doubling time of 144 h (at 7 °C), what gives a growth rate of $\mu = 0.12d^{-1}$, comparable to values for organisms listed in the Table 5.3. *Desulfobulbus rhabdoformis* was isolated from a water-oil separation system on a North Sea separation system and is an example of an anaerobic mesophilic propionate oxidizer and sulfate reducer with a doubling time $t_d = 11h$ ($\mu = 0.46d^{-1}$) on propionate and lactate in the presence of SO_4^{2-} (at 30 °C) (Lien et al. 1998). Nevertheless, there is still little known about acetogenic halophilic bacteria (Kivistö and Karp 2011), what, except for research on organisms growing under extreme conditions, makes further experiments on the culture enriched during this study highly interesting.

Obtained enrichment cultures of high purity are interesting in the view of their detailed description, as these might contain representatives of new propionate degrading species. During the research within the project on propionate degradation in different digesters regimes, the isolation of pure cultures was not successful; however, there is a follow up project within which further research on the cultures is undertaken. The direct aim of this study was not the pure cultures isolation, but getting a better insight in propionate degrading consortia, which was accomplished with studies on highly advanced enrichment cultures.

5.3.1 Propionic acid degradation rates

Organisms from different habitats analyzed during this study had propionate degradation rates between $0.004\,\mathrm{mmol\,l^{-1}\,h^{-1}}$ and $0.18\,\mathrm{mmol\,l^{-1}\,h^{-1}}$ (Table 5.3). The slowest rate was obtained in the sample enriched from North Sea sediments on propionate and sulfate. That was not expected, as sulfate is thought to be present in marine sediments at high concentrations, and should enhance propionate degradation in comparison to the approach without sulfate, which was ten times faster. However, other culture of similar origin without sulfate had a degradation rate of the same magnitude ($0.007\,\mathrm{mmol\,l^{-1}\,h^{-1}}$) (Table 5.3). In this case the temperatures of incubation were different, what could have an influence on differences between the degradation rate values. It is generally thought, from the thermodynamic point of view, that increasing temperature should increase the reaction rate (as it increases e.g. the diffusion constant of hydrogen/ formate and hence the flux, and the Gibbs free energy at pH = 7 and 55 °C of propionate degradation decreases to 62.3 kJ) and thereby also the degradation rate. Nevertheless the assumption from comparisons in Table 5.3 is opposite; the most effective cultures originate from mesophilic habitats. That accounts for both, the enrichment cultures and rates determined for non-enriched sludge. The only exception is a thermophilic lab-scale downflow packed bed reactor, where the highest rate among all compared propionate degraders was obtained. The fact that the propionate oxidation rate not only depended on temperature, but also on the type of feed or bacterial consortium pre-treatment cannot be forgotten in these considerations. Aguilar et al. (1995) for instance obtained better results in terms of propionate degradation in the culture that was pre-grown on glucose than on acetate, as the glucose pre-grown culture was more likely to contain propionate degraders due to the metabolites that were produced during digestion of this hexose. It is also hard to compare the rumen liquid capability of propionic acid degradation, as the rumen stomach is not a closed system in contrary to batch experiment, but it has been proved, that some rumen microorganisms are capable of propionate degradation with a higher rate than halophilic ones growing on propionate and sulfate, but lower than any other analyzed culture during this study. The culture originating from the CSTR digester ("tri-culture"), growing on propionate and sulfate was the "fastest" one obtained during the research

$(0.18\,\mathrm{mmol\,l^{-1}\,h^{-1}})$, and also very advanced in terms of enrichment. It contained three morphologically different microbial species and comparison of the degradation rate obtained for these organisms with other rates from Table 5.3, showed its leading position. That predestinates this culture to be applied as a healing agent (through bioaugmentation) for digesters during stagnation phases caused by VFAs accumulation.

Microorganisms' origin	Sample description	Propionic acid degradation rate	Reference
Lab-scale continuous **mesophilic** digester	Methanogenic culture pre-grown on acetic acid (batch assay)	$0.03\,\mathrm{mmol\,l^{-1}\,h^{-1}}$	Aguilar et al. 1995
Lab-scale continuous **mesophilic** digester	Methanogenic culture pre-grown on glucose (batch assay)	$0.18\,\mathrm{mmol\,l^{-1}\,h^{-1}}$	Aguilar et al. 1995
Industrial digester treating garbage, **thermophlic**	Lab-scale down-flow packed bed reactor	$7.71\,\mathrm{mmol\,l^{-1}\,h^{-1}}$ (maximum)	Tatara et al. 2008
Industrial biowaste digester, **mesophilic**	$900\,\mathrm{m^3}$ CSTR	$0.34\,\mathrm{mmol\,l^{-1}\,h^{-1}}$	Gallert and Winter 2008
Mesophilic, biowaste suspension	Lab-scale digester	$3.47\,\mathrm{mmol\,l^{-1}\,h^{-1}}$	Gallert et al. 2003
Mesophilic sewage sludge digester (sewage disposal plant)	Lab-scale digester with well acclimatized seed sludge	$0.20\,\mathrm{mmol\,l^{-1}\,h^{-1}}$	Wang et al. 1999
Mesophilic co-culture	*Syntrophobacter fumaroxidans* and *Methanobacterium formicicum*	$0.05\,\mathrm{mmol\,l^{-1}\,h^{-1}}$	Botsch and Conrad 2011
Water-oil separation system on a North Sea oil platform	*Desulfobulbus rhabdoformis* **mesophilic** growth on propionate + SO_4^{2-}	$\approx 0.33\,\mathrm{mmol\,l^{-1}\,h^{-1}}$	Lien et al. 1998
German Baltic Sea sediments	Sediments incubated at $12\,^\circ\mathrm{C}$	$0.007\,\mathrm{mmol\,l^{-1}\,h^{-1}}$	Boschker et al. 2001
$900\,\mathrm{m^3}$ **mesophilic** CSTR	Enrichment culture on propionate (8 transfers)	$0.12\,\mathrm{mmol\,l^{-1}\,h^{-1}}$	This study
Lab-scale digester treating market waste ITB, **mesophilic**	Enrichment culture on propionate (after 4 transfers)	$0.06\,\mathrm{mmol\,l^{-1}\,h^{-1}}$	This study

Lab-scale digester treating market waste LIPI, **mesophilic**	Enrichment culture on propionate (after 4 transfers)	$0.11\,\mathrm{mmol\,l^{-1}\,h^{-1}}$ This study
$900\,\mathrm{m^3}$ **mesophilic** CSTR	Enrichment culture on propionate + SO_4^{2-}	$0.18\,\mathrm{mmol\,l^{-1}\,h^{-1}}$ This study
Industrial scale **thermophilic** dry reactor treating biowaste	Enrichment culture on propionate + SO_4^{2-}	$0.06\,\mathrm{mmol\,l^{-1}\,h^{-1}}$ This study
Industrial scale **thermophilic** dry reactor treating biowaste	Enrichment culture on propionate	$0.06\,\mathrm{mmol\,l^{-1}\,h^{-1}}$ This study
Rumen liquid **(mesophilic)**	Rumen liquid suspended in mineral medium	$0.02\,\mathrm{mmol\,l^{-1}\,h^{-1}}$ This study
North Sea sediments (20 °C)	Suspension in a mineral medium with salt addition	$0.04\,\mathrm{mmol\,l^{-1}\,h^{-1}}$ This study
North Sea sediments (20 °C)	Enrichment on propionate and SO_4^{2-}	$0.004\,\mathrm{mmol\,l^{-1}\,h^{-1}}$ This study

Table 5.3: Propionic acid degradation rates for organisms of different origin.

5.4 Interspecies distances between propionic acid degraders and their partners

5.4.1 Propionic acid oxidation and growth rates

The propionate degrading enrichment culture originating from CSTR contained propionate degraders and hydrogenotrophic methanogens (after 8th transfer) which multiplied with a growth rate of $\mu = 0.062\,\mathrm{d^{-1}}$ (Figure 4.55). This value was determined from the increase of the total protein in the exponential growth phase (Figure 4.55). According to data from literature, that are compiled in Table 5.4, the growth rates obtained from different pure cultures of mesophilic syntrophic propionate degraders in co-culture with hydrogenotrophic methanogens were in the range from $0.066\,\mathrm{d^{-1}}$ to $0.19\,\mathrm{d^{-1}}$. Under sulfate-reducing conditions the growth rates were reaching values from $0.024\,\mathrm{d^{-1}}$ to $0.069\,\mathrm{d^{-1}}$ (Table 5.4). The growth rates of syntrophic propionate degraders in axenic cultures in the presence of sulfate are still much lower than those of the sulfate reducers alone in the presence of sulfate. This might mean that synthrophic acetogenic bacteria were possibly suppressed by sulfate reducers in sulfate-rich ecosystems.

Propionate degrader	Syntrophic partner/ Co-substrate	Growth rate μ [d^{-1}]	Reference
Pure cultures of syntrophic propionate-degrading bacteria			
Syntrophobacter fumaroxidans	*Methanospirillum hungatei*	0.17	Harmsen et al. 1998
	Fumarate (no propionate	) 0.09	
	Sulfate	0.024	Van Kujik and Stams 1995; Harmsen et al. 1998
Syntrophobacter pfennigii	*Methanospirillum hungatei*	0.066[1]	Wallrabenstein et al. 1995
	Sulfate	0.069[1]	
Syntrophobacter sulphatireducens	*Methanospirillum hungatei*	0.12	Chen at al. 2005
Syntrophobacter wolinii	*Methanospirillum hungatei* and *Desulfovibrio*	0.10[1]	Boone and Bryant 1980
	Sulfate and *Desulfovibrio*	0.19[1]	
	Sulfate	0.062	Wallrabenstein et al. 1994
Smithella propionica	*Methanospirillum hungatei*	n.d.[2]	Liu et al. 1999
Pelotomaculum schinkii	*Methanospirillum hungatei*; *Methanobacterium formicium*	0.099[1]	De Bok et al. 2005
Pelotomaculum thermopropionicum	*Methanothermobacter thermoautotrophicus*	0.19	Imachi et al. 2000; Imachi et al. 2002
	Pyruvate (no propionate)	1.65	
Desulfotomaculum thermocisternum	Hydrogenotrophic methane bacterium; sulfate	n.d.[2]	Nilsen et al. 1996
Desulfotomaculum thermobenzoicum subsp. thermosyntrophicum	*Methanobacterium thermoautotrophicum*	0.12	Plugge et al. 2002
	Pyruvate (no propionate)	0.33	
	Sulfate	0.099	
Pure cultures of selected sulfate reducing propionate degrading bacteria			

Desulfobulbus propionicus	Sulfate	0.89	Stams et al. 1984
	Sulfate	$1.66^{1)}$	Widdel and Pfenning 1982
Desulfobulbus elongatus	Sulfate	1.39	Samain et al. 1984
Desulforhabdus amnigenus	Sulfate	$0.46^{1)}$ (minimum)	Elferink et al. 1995
Strain PW	Methanobacterium formicium	$0.14^{1)}$	Wu et al. 1992
	Sulfate	$0.23^{1)}$	

[1] conversion of t_d into μ using $\mu = \ln 2 / t_d$

[2] not determined

Table 5.4: The comparison of growth rates for different propionic acid degrading bacteria.

The examined propionate degrading enrichment culture could oxidize propionate with a rate of $0.755 \, \mathrm{mmol\,l^{-1}\,d^{-1}}$ (Figure 4.60 a). This rate was higher than that for pure cultures (Table 5.4), but one order of magnitude lower than the rate of propionate oxidation in the biowaste CSTR full scale mesophilic digester for treatment of a biowaste suspension from the same origin ($8.10 \, \mathrm{mmol\,l^{-1}\,d^{-1}}$, Gallert and Winter 2008). It was two orders of magnitude lower than the rate of propionate oxidation by mesophilic sludge of a lab scale digester, also treating a biowaste suspension ($83.2 \, \mathrm{mmol\,l^{-1}\,d^{-1}}$, Gallert et al. 2003). Since conversion rates in general depend on the biomass concentration, it is necessary to relate degradation to active biomass, volatile suspended solids (VSS) or protein. To allow comparison of obtained results with data from literature the specific degradation rates were calculated from the total protein content under the assumption, that 50 % of the dry matter of bacteria would be protein. Thus, the specific propionate oxidation rate of 49 mmol per (mg protein) per day (Figure 4.60 b) of the enrichment culture corresponding to 98 mmol per (mg VSS) per day. This propionate oxidation rate is much higher than the propionate oxidation rate of 1.1 mmol per (g VSS) per day measured in the sludge from an anaerobic hybrid reactor (Lens et al. 1996) or of 9.14 mmol per (mg VSS) per day measured in granular sludge (Schmidt and Ahring 1995).

5.4.2 Acetic acid inhibition

The degradation of propionate was inhibited more than 50 % if acetate accumulated to 20 mmol l^{-1} (Figure 4.61). Similar results were obtained by Mawson et al. (1991) who reported a 50 % inhibition of propionate oxidation in the presence of 33 mmol l^{-1} acetate. This was not caused by a pH-change, since the pH of the medium was sufficiently buffered. Additionally, Gorris et al. (1989) determined almost complete inhibition of propionate degradation in a mesophilic lab-scale digester during start-up at acetate concentrations exceeding 8 mmol l^{-1}. Acetic acid present at levels higher than 23 mmol l^{-1} also inhibited propionic acid degradation in a mesophilic lab-scale digester seeded with sewage sludge (Wang et al. 1999).

5.4.3 Propionic acid degraders grow at the energetical limit

Propionate degraders have to live with the "smallest quantum of energy in biology" (Müller et al. 2010) available for their growth. Members of the propionate degrading enrichment culture (Figure 4.55), where most syntrophs degrading this acid were related to *S. fumaroxidans*, reached the growth yield equal to 4.1 g biomass per mol propionate (20.3 mmol l^{-1} propionate oxidation resulted with 41.8 mg protein production, protein equals 50 % dry mass). A maximum biomass yield of 2.6 g mol^{-1} propionate was previously reported by Scholten and Conrad (2000) during the experiments on *S. fumaroxidans* and *M. hungatei* in a co-culture. An increase up to 5.6 g mol^{-1} propionate was reached in chemostat cultures, which were flushed with nitrogen for more efficient hydrogen removal. Most of the propionate degraders use the methyl-malonyl-CoA pathway (Plugge et al.1993) for propionate degradation where the yield of 1 mol ATP per mol propionate is produced via substrate level phosphorylation. In the presence of hydrogen consumers such as hydrogenolytic methanogens, the hydrogen partial pressure can be maintained low enough to reduce protons directly with electrons from the oxidation of malate to oxaloacetate or of pyruvate to acetyl-CoA, but not with electrons from the oxidation of succinate to fumarate (Plugge et al. 1993, Müller et al. 2010). Reversed electron transport, driven by hydrolysis of 2/3 mol ATP would make succinate oxidation possible (Schink 1997, Schink and Stams 2006) and "only" 1/3 mol ATP would then be available for growth. Generally, per 1 mol ATP, 5 to 12 g biomass can be produced by bacteria during fermentation (Stouthamer 1979). Resulting 4.1 g biomass per mol propionate or 1/3 mol ATP would fit in the upper limit of 12 g biomass per mol ATP, what designates efficient conversion into biomass of the examined enrichment culture.

5.4.4 Hydrogen and formate as electron carriers

Hydrogen or formate could be electron carriers from producers (acetogenic bacteria) to consumers (methanogenic bacteria) and the efficiency of electron transfer depends on interspecies distances. Hydrogen has a higher diffusivity than formate, but formate is more soluble than hydrogen. Therefore, hydrogen transfer would be advantageous at low interspecies distances and formate transfer would be advantageous over longer interspecies distances (de Bok et al. 2004). Possibilities to reduce transfer distances in liquid suspensions include either the use of artificially concentrated, dense cultures or of autoaggregated flocs, granules or pellets. Floc formation by the examined enrichment culture from a biowaste digester during syntrophic propionate degradation was observed (Figure 4.56 c - j) and an increasing size of the flocs up to 52 µm was determined (Figure 4.58). Cluster formation of fatty acid-oxidizing acetogenic bacteria and H_2-oxidizing methanogenic bacteria seems to facilitate hydrogen transfer (Conrad et al. 1986). Microscopic observations of flocs in the syntrophic methanogenic cultures revealed the close neighborhood of acetogens and methanogens, allowing conclusions on the kinetics of interspecies electron transfer within the flocs (Thiele et al. 1988). The close neighborhood of propionate degraders and of methanogens in floc-forming enrichment culture examined during this study can be seen in Figure 4.56 c - j. In older cultures, a few micro-colonies of just propionate-degraders can also be seen in the flocs. Micro-colony formation in granular sludge flocs, fed with propionate were also observed by Harmsen et al. (1996). Structured layers in granules with a thick outer layer almost free of methanogens and layers below, containing microcolonies of both methanogens and propionate-degraders, could be visualized. In contrary, microcolonies of only one trophic group of bacteria were observed in the flocs of a whey-degrading chemostat (Thiele et al. 1988). The advantage of aggregation in syntrophic populations that degraded volatile fatty acids via interspecies hydrogen transfer was also investigated in sludge granules. With propionate as substrate a 50 % higher H_2 consumption, measured as specific methanogenic activity (SMA), and a 20 % higher propionate oxidation was reported for intact granular sludge, as compared to bacterial suspensions that were prepared by disintegration of the granules (Shi-yi and Jian 1992, Schmidt and Ahring 1995). In experiments with co-cultures of thermophilic *Pelotomaculum thermopropionicum* and *Methanobacterium thermoautotrophicum* co-aggregation apparently depended on the type of substrate. With thermodynamically unfavorable propionate, a large fraction of cells co-aggregated, whereas a relatively small proportion of cells co-aggregated with ethanol or 1-propanol as substrates (Ishii et al.

2005). In suspended propionate-degrading syntrophic cultures, one option to decrease interspecies distances is to increase the proportion of hydrogenotrophic methanogens for enhanced mass transfer of metabolites (Amani et al. 2012).

The methanogens in examined propionate-degrading enrichment culture were able to produce methane from H_2/CO_2 and at a much lower rate from formate (Figure 4.62 a). Vice versa formate was generated by the methanogens when H_2/CO_2 was introduced into the gas phase and thus surplus hydrogen was available (Figure 4.62 b). Some formate (maximally 0.09 mmol l^{-1}) was also excreted by this culture in the presence of propionate when hydrogenothrophic methanogenesis was completely inhibited. Formate formation from H_2/CO_2 would only require an active formate dehydrogenase and might stem from the methanogens, the propionate-degraders or from both. With fumarate as an electron acceptor Dong and Stams (1995) reported a minimal concentration of 0.024 mmol l^{-1} formate that was generated by the MPOB's from propionate with fumarate as electron acceptor. In this study, the non-BESA-inhibited batch assay with propionate no formate was detected. In the literature a controversial discussion is documented about the requirement or participation of formate, hydrogen or both for electron interspecies transfer. Some authors state that both are equally important for syntrophic bacteria (Batstone et al. 2006, Schink 1997) or that formate may be more significant (de Bok et al. 2004) with fluxes a 100 times higher than those for hydrogen in a co-culture of MPOBs and *M. hungatei* (Dong and Stams, 1995). Formate as an electron shuttle component may not be significant in suspensions of disintegrated thermophilic granules (Schmidt and Ahring, 1993) or for mesophilic granular sludge (Schmidt and Ahring 1995).

5.5 Microbial organization within the aggregates

The spatial distribution of propionate degraders and methanogens in examined syntrophic enrichment culture was visualized with fluorescence *in - situ* hybridization using specific 16S rRNA-targeted oligonucleotides labeled with FAM (propionate-degraders) or Cy3 (methanogenic bacteria). The species of both trophic groups of bacteria in the flocs were almost randomly distributed with the formation of only very little micro-colonies. In order to quantify population changes during the different growth phases the daime software for digital image analysis in microbial ecology was implemented (Daims et al. 2006) and biovolume fractions were determined. The total biovolume was obtained from DAPI-stained preparations, whereas the biovolume fractions of the propionate-degrading acetogens and the methane-forming archaea were determined with group-specific 16S rRNA-based oligonucleotide probes. Although DAPI has its limitations for such kind of evaluation (Nielsen et al. 2009), using it as a basis for the description of population fluctuations at each growth phase gave a sufficiently reliable "external" marker for changes of the ratio of propionate-degraders and methanogens. The relative abundance of methanogens that hybridized with the specific 16S rRNA-targeted oligonucleotide probe Arc 915 remained almost constant over the whole period of 42 days with 4 propionate feedings and accounted for approximately 52 % of the total biovolume in relation to unspecific DAPI staining (Figure 4.57). Comparable results were reported for granular sludge fed with propionate, where the archaeal population (16S rRNA) remained constant at 55 % over 12 weeks (Harmsen et al. 1996). Park et al. (2009) also reported the fraction of methaogens (FISH, Arc 915) to remain between 49 and 54 % in enrichment cultures on propionate originating from a sewage treatment plant for 26th to 68th day of growth. The biovolume fraction of propionate degraders in the examined enrichment culture, that hybridized with the 16S rRNA-targeted oligonucleotide probe KOP1 (*"Syntrophobacter pfennigii"*-related organisms) and MPOB1 (*"Syntrophobacter fumaroxidans"*-related organisms, including also *"Syntrophobacter pfennigii"*-related organisms), increased with time during the different growth phases. The biovolume fraction of MPOB-like propionate degraders accounted for 50 % after 42 days and was almost identical with the percentage-value for the biovolume fraction that was obtained after hybridization with the universal EUB-probe (Figure 4.57). Resilience and resistance to e.g. elevated hydrogen partial pressure of both, methanogens and syntrophic propionate degraders must guarantee a stable metabolic function over time for survival, despite of disturbances.

Aggregate formation was a possibility to decrease the interbacterial distance, which was necessary for a fast transfer of metabolites. In the literature interspecies distances of 12.4 µm in co-cultures of *S. fumaroxidans* and *M. hungatei* (Stams and Dong 1995), of 4 µm in mesophilic granular sludge between hydrogen producing and hydrogen consuming bacteria (Stams et al. 1989), of 2 µm to 3 µm in intact mesophilic sludge granules (Schmidt and Ahring 1995) or of 2.3 µm in aggregates of thermophilic *P. thermopropionicum* and *M. thermoautotrophicum* (Ishii et al. 2005) were reported. In our experiments, the interspecies distances decreased from 5.3 µm in the cell suspension directly after inoculation to 1.56 µm in phase 1, where small flocs were dominant, to 0.29 µm in phase 4 with mainly larger flocs (Figure 4.58). According to Fick's diffusion law (1.8) an inverse relationship between syntrophic distance and hydrogen flux can be seen. A low spatial distance with a high diffusion rate would allow maintenance of a low hydrogen partial pressure and enhance syntrophic propionate degradation. Calculation of the hydrogen concentrations according to (4.5) gave minimum H_2 concentrations of 0.32 Pa for the methanogens and maximum H_2 concentrations of 9.34 Pa for the propionate-oxidizers delivering a maximum possible $C_{PA} - C_{H_2}$ value. According to de Bok et al. (2004), highest H_2 levels formed by *S. fumaroxidans* were 6.8 Pa and lowest H_2 level reached by *M. hungatei* were 2 Pa. In similar experiments of Scholten and Conrad (2000) with a syntrophic co-culture of *S. fumaroxidans* and *M. hungatei*, maximum H_2 concentrations of 9 Pa and minimum H_2 concentrations of 0.8 Pa, respectively, were measured. These values were in accordance with those calculated for our co-culture experiments. The maximum possible hydrogen flux J_{H_2} in the successive growth phases increased with decreasing interspecies distances. The maximum possible J_{H_2} in phase 1 was 1.1 nmol H_2 /ml/min, which increased to 10.3 nmol H_2 /ml/min in phase 4 (Table 4.10). A hydrogen flux of 1.1 nmol H_2 /ml/min in co-cultures of the propionate-degrading strain MPOB and *M. hungatei* was reported (Stams and Dong 1995) or of 5.38 nmol H_2 /ml/min in co-cultures of *P. thermopropionicum* and *M. thermoautotrophicus* can be calculated from the data of Ishii et al. (2005).

Some computer-based simulation experiments for syntrophic degradation with hydrogen transfer resulted in co-location of hydrogen producers and hydrogen consumers, caused by high uptake rates of hydrogen consumers rather than sensitivity of hydrogen producers to hydrogen inhibition (Batstone et al. 2006). Results with the examined enrichment culture indicated, that methanogens and syntrophic propionate degraders "find each other" after inoculation and form aggregates. By lowering the interspecies distance the hydrogen flux from the hydrogen producer to the hydrogen consumer will not limit the overall syntrophic interaction.

5.6 Optimal conditions for growth of propionate degraders

Research with various microbial consortia degrading propionic acid done during this study shows that it is hard to find the optimal parameters for the growth of propionate degraders. From six mesophilic samples obtained from Indonesia for metabolic pathway determination (Table 4.6), only three could degrade propionic acid at pH $= 7.0$ and 37 °C. Enrichment cultures on propionate with microorganisms originating from LIPI and ITB lab-scale digesters treating market waste were morphologically different (Figure 4.32), and hence degraded propionate at different rates (Table 5.3). Enhanced degradation of this acid was observed in the sample containing Methanosarcina sp. (LIPI), responsible for acetate removal. The ITB enrichment sample was growing with slight acetate production, what implies acetate degraders' presence, however the degradation rate of propionate was significantly slower. In both of them the aggregation of microorganisms was noticed (Figure 4.32). The thermophilic enrichments from dry fermentation obtained by means of pasteurization were relatively slow, consisting of two morphologically different organisms which were not forming aggregates (Figure 4.54), and among which no acetate degraders were present. The number of organisms in the enrichment culture on propionate originating from rumen liquid was quite low, the degradation rate was very low (but still higher than in a halophilic culture), organisms formed small aggregates (Figure 4.46) and acetate degradation was recorded (Figure 4.45). Halophilic microorganisms grew on propionate with lowest degradation rate. They required CO_2 for growth (Figure 4.48), and did not produce measurable concentrations of acetate (Figure 4.48). Additionally aggregates' formation was observed (Figure 4.50). The best results were obtained for the mesophilic enrichment culture from an industrial CSTR digester treating biowaste. The enrichment culture containing three morphologically different organisms was capable of forming aggregates (Figure 4.35) during growth on propionate under sulfate reduction and acetate production (Figure 4.37).

Complete mineralization of propionate seems to be possible by three different microorganisms'- propionate oxidizers, hydrogen/ formate consumers and acetate degraders (Stams 1994). That should imply, that optimal propionate degradation can be maintained, when the three partnering bacteria cooperate. This study shows that it might be true, as acetic acid inhibits propionic acid degradation (Figure 4.61). This goes along with calculations done by Stams et al. (1989), which confirm the dependence of propionate degradation on hydrogen and acetate removal from the system. Additionally, Stams et al. (1992) reported that a co-culture of a thermophilic propionate degrader with *M.thermoautotrophicum* had a specific growth rate between $0.07\,d^{-1}$ to $0.19\,d^{-1}$, which was lower than the specific growth rate of - $0.25\,d^{-1}$ to $0.32\,d^{-1}$ determined for their tri-culture where both, hydrogenotrophic and aceticlastic methanogens (*M. thermautotrophicum* and *M. thermophila*) were present.

Schink and Stams (2006) considered *Methanospirillum hungatei* as improper choice for investigations on aggregation mechanisms, as this methanogen does not enhance microbial aggregation. The microscopic images of *S. pfen-*

nigii and *M. hungatei* co-cultures done by Wallrabenstein et al. (1995) or *S. wolinii* and *M. hungatei* (Wallrabenstein et al. 1994) might serve as a proof for that. The thermophilic culture isolated during this research also did not form aggregates, what might be caused by the methanogenic partner, as it was morphologically similar to *M. hungatei*. During this studiy, aggregation occurred between non-filamentous methane bacteria and MPOB-like propionate degraders, and that allows using it as a test culture for aggregation mechanism determination. That would be consistent with findings of Botsch and Conrad (2011) where *S. fumaroxidans* formed aggregates during growth in a co-culture with Methanobacterium formicicum on propionate. In case of *Pelotomaculum schinkii*, there was no difference in growth rate between co-cultures with *M. hungatei* or *M. formicium* (Table 5.4). This might indicate that there are no contradictions with *M. formicicum* co-culturing with propionate degraders.

During the growth of strain SI on propionate in a co-culture with *M. thermoautotrophicum* some instability was noticed by Imachi et al. (2000) due to mechanical disruption of spatial microbial distribution (extending the distances between propionate degrader and methanogenic partner). Similarly, Stams et al. (1992) found that shaking a low-density co-culture of *Methanobacterium thermoautotrophicum* with propionate degraders enriched from the granular methanogenic sludge of a bench-scale upflow anaerobic sludge bed reactor caused starvation of the methanogens, as they could not get enough hydrogen for growth. They concluded that influencing the close positioning of participants significantly enhances the degradation rates of propionate. The findings from this study confirm above discussion, as the enrichment cultures in which organisms co-aggregated could degrade propionate with a higher rate than those growing in a free suspension (thermophilic cultue). And that despite the fact, that all enrichments were cultivated without shaking to enhance the aggregation. There was always some fraction of organisms that were not attached to aggregates (e.g. Figure 4.56 h), but they were not the focus of this research. Many researchers report finding propionate degraders in a biofilm or a granule of anaerobic reactors (e.g. Tatara et al. 2008, Zellner and Neudörfer 1995, Zheng et al. 2006). The behavior of a "suspende"? biomass, however, e.g. after the bioaugmentation with propionate degrading enrichment (tri-culture) might be an interesting issue. As well as estimating the ratio of bacteria remaining in free-suspension to the aggregating ones in co-cultures and comparison of results with microbial behavior in digesters.

All in all, it appears that propionate degradation in anaerobic digesters proceeds within aggregates (this study), or, according to literature discussed before, within granules and biofilm layers, as this is the localization where syntrophs are most abundant and where they are not directly exposed to high hydrogen partial pressure, which might be present in the reactor. Additionally, the studies on propionate degradation inhibition by adding hydrogen gas to mixed cultures of methanogenic bacteria immobilized in porous polyurethane particles in extended fixed bed reactor did not cause any observable effect on propionic acid removal (Kus and Wiesmann 1995) which seem to support this thesis. That was explained with a fact, that hydrogen was consumed by a thin layer of the organisms on the particle with high numbers of hydrogen consumers.

There are several factors that have to be maintained for optimal propionate degradation and balancing them is not straightforward. In terms of propionic acid conversion into methane, acetic acid, hydrogen and/ or formate have to be efficiently removed from the system. The efficiency of this phenomenon is related to spatial distances, cooperating partners or reducing agents availability and their form of appearance like e.g. aggregate or granule.

Bibliography

Abalos M. and Bayona J.M. (2000) Application of gas chromatography coupled to chemical ionisation mass spectrometry following headspace solid-phase microextraction for the determination of free volatile fatty acids in aqueous samples. *J Chromatogr A*, 89:287–294.

Abildgraad L, Ramsing N.B., and Finster K. (2004) Characterization of the marine propionate-degrading, sulfate reducing bacteria *Desulfobaba fastidiosa* sp. nov. and reclassification of *Desulfomusa hansenii* as *Desulfobaba hansenii* comb. nov. *Int J Syst Evol Microbiol*, 54:393–399.

Aguilar A., Casas C., and Lema J.M. (1995) Degradation of volatile fatty acids by differently enriched methanogenic cultures: kinetics and inhibition. *Water Res*, 29:505–509.

Amani T., Nosrati M., and Mousavi S.M. (2012) Response surface methodology analysis of anaerobic syntrophic degradation of volatile fatty acids in an upflow anaerobic sludge bed reactor inoculated with enriched cultures. *Biotechnol Bioprocess Eng*, 17:133–144.

Amann R.I., Krumholz L., and Stahl D.A. (1990) Fluorescent-oligonucleotide probing of whole cells for determinative phylogentic, and environmental studies in microbiology. *J Bacteriol*, 172:762–770.

Amman R.I., Binder B.J., Olson R.J., Chisholm S.W., Devereux R., and Stahl D.A. (1990) Combination of 16S rRNA-targeted oligonucleotide probes with flow cytometry for analyzing mixed microbial population. *Appl Environ Microbiol*, 56:1919–1925.

Angenent L.T., Sung S., and Raskin L. (2002) Methanogenic population dynamics during start-up of a full-scale anaerobic sequencing batch reactor treating swine waste. *Water Res*, 36:4648–4654.

Appels L., Lauwers J., Degreve J., Helsen L., Lievens B., Willems K., van Impe J., and Dewil R. (2011) Anaerobic digestion in global bio-energy production: potential and research challenges. *Renew Sustain Energy Rev*, 15:4295–4301.

Ariesyady H.D., Ito T., and Yoshiguchi K. (2007) Phylogenetic and functional diversity of propionate-oxidizing bacteria in anaerobic digester sludge. *Appl Microbiol Biotechnol*, 75:673–683.

Asinari Di San Marzano C.M., Binot R., Bol T., Fripiat J., Hutschemakers J., Melchior J.L., Perez I., Nveau H., and Nyns E.J. (1981) Voaltile fatty acids, an important state parameter for the control of the reliability and the productivities of methane anaerobic digestions. *Biomass*, 1:47–59.

Balch W.E., Fox G.E., Magrum L.J., Woese C.R., and Wolfe R.S. (1979) Methanogens reevaluation of a unique biological group. *Microbiol Rev*, 43:260–296.

Banel A. (2010) *Opracowywanie i porównanie chromatograficznych metodyk oznaczania krótkołańcuchowych kwasów alkanomonokarboksylowych w próbkach wodnych i stałych*. PhD thesis, Gdańsk University of Technology.

Banel A., Wasielewska M., and Zygmunt B. (2011) Application of headspace solid-phase microextraction followed by gas chromatography-mass spectrometry to determine short-chain alkane monocarboxylic acids in aqueous samples. *Anal Bioanal Chem*, 399:3299–3303.

Barredo M.S. and Evison L.M. (1991) Effect of propionate toxicity on methanogenic-enriched sludge, *Methanobrevibacter smithii*, and *Methanospirillium hungatei* at different pH values. *Appl Environ Microbiol*, 57:1764–1769.

Bass O., Lascourreges J.F., Le Borgne F., Le Goff C., and Magot M. (2009) Characterization by culture and molecular analysis of the microbial diversity of a deep subsurface gas storage aquifer. *Res Microbiol*, 160:107–116.

Batstone D.J., Picioreanu C., and van Loosdrecht M.C.M. (2006) Multidimensional modeling to investigate interspecies hydrogen transfer in anaerobic biofilms. *Water Res*, 40:3099–3108.

Beaty P.S. and McInerney M.J. (1987) Growth of *Syntrophomonas wolfei* in pure cultures on crotonate. *Arch Microbiol*, 147:389–393.

Bleicher K. and Winter J. (1994) Formate production and utilization by methanogens and by sewage sludge consortia - Interference with the concept of interspecies formate transfer. *Appl Microbiol Biotechnol*, 40:910–915.

Boe K., Batstone D.J., and Angelidaki I. (2007) An innovative online VFA monitoring system for the anaerobic process, based on headspace gas chromatography. *Biotechnol Bioeng*, 96:712–721.

Boone D.R. and Bryant M.P. (1980) Propionate-degrading bacterium, *Syntrophobacter wolinii* sp. Nov. gen. nov., from methanogenic ecosystems. *Appl Environ Microbiol*, 40:626–632.

Boone D.R., Johnson R., and Liu Y. (1989) Diffusion of the interspecies electron carriers H_2 and formate in methanogenic ecosystems and its implications in the measurement of Km for H_2 or formate uptake. *Appl Environ Microbiol*, 55:1735–1741.

Boone D.R., Whitman W.B., and Rouviere P. (1993) *Methanogenesis: Ecology, physiology, biochemistry and genetics* . New York: Chapman & Hall. Diversity and taxonomy of methanogens.

Borja R. (2011) *Comprehensive Biotechnology*. Burlington: Academic Press, second edition edition. Biogas production.

Boschker H.T.S., De Graaf W., Köster M., Meyer-Reil L.A., and Cappenberg T.E. (2001) Bacterial populations and process involved in acetate and propionate consumption in anoxic brakish sediment. *FEMS Microbiol Ecol*, 35:97–103.

Botsch K.C. and Conrad R. (2011) Fractionation of stable carbon isotopes during anaerobic production and degradation of propionate in defined microbial cultures. *Organic Geochem*, 42:289–296.

Bowles M.W., Samarkin V.A., Bowles K.M., and Joye S.B. (2011) Weak coupling between sulfate reduction and the anaerobic oxidation of methane in methane-rich seafloor sediments during *ex situ* incubation. *Geochimica Cosmochimica Acta*, 15:500–519.

Briones A.M., Daugherty B.J., Angenent L.T., Rausch K., Tumbleson M., and Raskin L. (2009) Characterization of microbial trophic structures of two anaerobic bioreactors processing sulfate-rich waste streams. *Water Res*, 43:4451–4460.

Brusa T., Del Puppo E., Ferrari A., Rodondi G., Andreis C., and Pellegrini S. (1997) Microbes in deep-sea anoxic basins. *Microbiol Res*, 152:45–56.

Bryant M.P. (1979) Microbial methane production- theoretical aspects. *J Animal Sci*, 48:193–201.

Bryant M.P., Wolin E.A., Wolin M.J., and Wolfe R.S. (1967) *"Methanobacillus omelanskii"*, a symbiotic association of two species of bacteria. *Arch Microbiol*, 59:20–31.

Chachikani M., Dabert P., Abzianidze T., Partskhaladze G., Tsiklauri L., Dudauri T., and Godon J.J. (2004) 16S rDNA characterisation of bacterial and archaeal communities during start-up of anaerobic thermophilic digestion of cattle manure. *Biores Technol*, 93:227–232.

Chandra R., Takeuchi H., and Hasegawa T. (2012) Methane production from lignocellulosic agricultural crop wastes: a review in context to second generation of biofuel production. *Renew Sustain Energy Rev*, 16:1462–1476.

Chen A., Liao P.H., and Lo K.V. (1994) Headspace analysis of malodorous compounds from swine wastewater under aerobic treatment. *Biores Technol*, 49:83–87.

Chen S., Liu X., and Dong X. (2005) *Syntrophobacter sulfatireducens* sp. nov., a novel syntrophic, propionate-oxidizing bacterium isolated from UASB reactors. *Int J Syst Evol Microbio*, 55:1319–1324.

Chen Y., Cheng J., and Creamer K. (2008) Inhibition of anaerobic digestion process: A review. *Biores Technol*, 99:4044–4064.

Chong S., Sen T.K., Kayaalp A., and Ang H.M. (2012) The performance enhancements for upflow anaerobic sludge blanket (UASB) reactors for domestic sludge treatment- a state-of-the-art review. *Water Res*, 46:3434–3470.

Conrad R., Schink B., and Phelps T.J. (1986) Thermodynamics of H_2-consuming and H_2-producing metabolic reactions in diverse methanogenic environments under *in situ* conditions. *FEMS Microbiol Ecol*, 38:353–360.

Cruvys J.A., Dinsdale R.M., Hawkes F.R., and Hawkes D.L. (2002) Development of a static headspace gas chromatographic procedure for the routine analysis of volatile fatty acids in wastewaters. *J Chromatogr A*, 945:195–209.

Dabert P., Delgenés J.P., Moletta R., and Godon J.J. (2002) Contribution of molecular microbiology to the study in water pollution removal of microbial community dynamics. *Rev Environ Sci Biotechnol*, 1:39–49.

Daims H. (2009) *FISH handbook for biological wastewater treatment: identification and quantification of microorganisms in active sludge and biofilms by FISH*. London: IWA Publishing. Quantitative FISH for the cultivation-independent quantification of microbes in wastewater treatment plants.

Daims H., Lücker S., and Wagner M. (2006) *daime*, a novel image analysis programe for microbial ecology and biofilm research. *Environ Microbiol*, 8:200–213.

de Bok F.A.M., Harmsen H.J.M., Plugge C.M., de Vries M.C., Akkermans D.L., de Vos W.M., and Stams A.J.M. (2005) The first true obligately syntrophic propionate-oxidizing bacterium, *Pelotomaculum schinkii* sp. nov., co-cultured with *Methanospirillum hungatei*, and emended description of the genus *Pelotomaculum*. *Int J Syst Evol Microbiol*, 55:1697–1703.

de Bok F.A.M., Plugge C.M., and Stams A.J.M. (2004) Interspecies electron transfer in methanogenic propionate degrading consortia. *Water Res*, 38:1368–1375.

de Bok F.A.M. and Roze E.H.A.and Stams A.J.M. (2002) Hydrogenases and formate dehydrogenases of *Syntrophobacter fumaroxidans*. *Antonie van Leeuwenhoek*, 81:283–291.

de Bok F.A.M., Stams A.J.M., Dijkema C., and Boone D.R. (2001) Pathway of propionate oxidation by syntrophic culture of *Smithella propionica* and *Methanospirillum hungatei*. *Appl Environ Microbiol*, 67:1800–1804.

Deublein D. and Steinhauser A. (2008) *Biogas from waste and renewable sources: an introduction*. Weinheim: Wiley-VCH Verlag GmbH & Co. KGaA.

Dhaked R.K., Singh P., and Singh L. (2010) Biomethanation under psychrophilic conditions. *Waste Manag*, 30:2490–2496.

Dincer F., Odabasi M., and Muezzinoglu A. (2006) Chemical characterization of odorous gases at a landfill site by gas chromatography-mass spectrometry. *J Chromatogr A*, 1122:222–229.

Dong X. and Stams A.J.M. (1995) Evidence for H_2 and formate formation during syntrophic butyrate and propionate degradation. *Anaerobe*, 1:35–39.

Dröge S., Ewen A., and B. Pacan (2008) Wenn es Spuren mangelt. *Biogas J*, 2:30–35.

Dwyer D.F., Weeg-Aerssens E., Shelton D.R., and J.M. Tiedje (1988) Bioenergetic conditions of butyrate metabolism by a syntrophic, anaerobic bacterium in coculture with hydrogen-oxidizing methanogenic and sulfidogenic bacteria. *Appl Environ Microbiol*, 54:1354–4359.

Elferink S.J.W.H.O., Maas R.N., Harmsen H.J.M., and Stams A.J.M. (1995) *Desulforhabdus amnigenus* gen. nov. sp. nov., a sulfate reducer isolated from anaerobic granular sludge. *Arch Microbiol*, 164:119–124.

Fall P.A.D. (2002) *FISH zur Überwachung von Biogasreaktoren*. PhD thesis, Technische Universität München.

Federal Ministry for the Environment, Nature Conservation and Nuclear Safety (2012) www.bmu.de. Accessed Nov. 8th, 2012.

Fernandez A.S., Hashsham S.A., Dollhopf S.L., Raskin L., Glagoleva O., Dazzo F.B., Hickey R.F., Criddle C.S., and Tiedje J.M. (2000) Flexible community structure correlates with stable community function in methanogenic bioreactor communities perturbed by glucose. *Appl Environ Microbiol*, 66:4058–4067.

Ferry J.G., Smith P.H., and Wolfe R.S. (1974) *Methanospirillum*, a new genus of methanogenic bacteria, and characterization of *Methanospirillum hungatii* sp.nov. *Int J Syst Bacteriol*, 24:465–469.

Finster K., Thomsen T., and Ramsing N.B. (2001) *Desulfomusa hansenii* gen. nov., sp. nov., a novel marine propionate-degrading, sulfate reducing bacterium isolated from *Zostera marina* roots. *Int J Syst Evol Microbiol*, 51:2055–2061.

Fox P. and Pohland F.G. (1994) Anaerobic treatment applications and fundamentals: substrate specifity during phase separation. *Water Environ Res*, 66:716–724.

Gallert C., Henning A., and Winter J. (2003) Scale-up of anaerobic digestion of the biowaste fraction from domestic wastes. *Water Res*, 37:1433–1441.

Gallert C. and Winter J. (2005) *Environmental Biotechnology - Concept and applications*. Weinheim: Wiley-VCH GmbH & Co. KGaA,. Bacterial metabolism in wastewater treatment systems.

Gallert C. and Winter J. (2008) Propionic acid accumulation and degradation during restart of a full-scale anaerobic biowaste digester. *Bioresour Technol*, 99:170–178.

Gerardi M.H. (2003) *The microbiology of anaerobic digesters*. New Jersey: John Wiley and Sons.

Gong P., Sun T.H., Beudert G., and Hahn H.H. (1997) Ecological effects of combined organic or inorganic pollution on soil microbial activities. *Water Air Soil Pollut*, 96:133–143.

Gonzalez-Fernandez C. and Garcia-Encina P.A. (2009) Impact of substrate to inoculum ratio in anaerobic digestion of swine slurry. *Biomass Bioeng*, 33:1065–1069.

Gorby Y.A., Yanina S., McLean J.S., Rosso K.M., Moyles D., Dohnalkova A., Beveridge T.J., Chang I.S., Kim K.S., Culley D.E., Reed S.B., Romine M.F., Saffarini D.A., and et.al. (2006) Electrically conductive bacterial nanowires produced by *Shewanella oneidensis* strain MR-1 and other microorganisms. *Proc Natl Acad Sci USA*, 103:11358–11363.

Gorris L.G.M., van Deursen J.M.A., van der Drift C., and Vogels G.D. (1989) Inhibition of propionate degradation by acetate in methanogenic fluidized bed reactors. *Biotechnol Lett*, 11:61–66.

Gujer W. and Zehnder A.J.B. (1983) Conversion processes in anaerobic digestion. *Water Sci Technol*, 15:127–167.

Hahn H.H. and Hoffmann E. (1992) Vergärung organischer Abfälle - getrennt oder gemeinsam mit Klärschlamm? In *Abfall ist kein Zufall - Umweltkongress Mannheim*, pp D1–D7.

Hajaranis S.R. and Ranade D.R. (1994) Inhibition of methanogens by n- and iso-volatile fatty acids. *World J Microbiol Biotechnol*, 10:350–351.

Harmsen H.J.M., Akkermans A.D.L., Stams A.J.M., and De Vos W.M. (1996) Population dynamics of propionate-oxidizing bacteria under methanogenic and sulfidogenic conditions in anaerobic granular sludge. *Appl Environ Microbiol*, 62:2163–2168.

Harmsen H.J.M., Kengen H.M.P., Akkermans A.D.L., and Stams A.J.M. (1995) Phylogenetic analysis of two syntrophic propionate-oxidizing bacteria in enrichment cultures. *Syst Appl Microbiol*, 18:67–73.

Harmsen H.J.M., Van Kuijk B.L.M., Plugge C.M., Akkermans A.D.L., De Vos W.M., and Stams A.J.M. (1998) *Syntrophobacter fumaroxidans* sp. nov., a syntrophic propionate-degrading sulfate-reducing bacterium. *Int J Syst Bacteriol*, 48:1383–1387.

Hatamoto M., Imachi H., Ohashi A., and Harada H. (2007) Diversity of anaerobic microorganisms involved in long-chain fatty acid degradation in methanogenic sludges as revealed by RNA-based stable isotope probing. *Appl Environ Microbiol*, 73:4119–4127.

Heitefuss S., Heine A., Horst S., and Seifert H. (1990) Detection of non-volatile organic acids by head-space gas chromatography. *J Chromatogr*, 532:374–378.

Himanen M., Latva-Kala K., Itävaara M., and Hänninen K. (2006) A method for measuring low-weight carboxylic acids from biosolid compost. *J Environ Qual*, 35:516–521.

Hines M.E., Evans R.S., Sharak Genthner B.R., Willis S.G., Friedman S., Rooney-Vagra J.N., and Devereux R. (1999) Molecular phylogenetic and biogeochemical studies of sulfate-reducing bacteria in the rhizosphere of *Spartina alterniflora*. *Appl Environ Microbiol*, 65:2209–2216.

Höpner T. and Knappe J. (1974) *Methoden der enzimatischen Analyse*, vol. 3. Veinheim: Chemie Verlag.

Hoshika Y. (1981) Gas-chromatographic determination of trace amounts of lower fatty acids in ambient air near and in exhaust gases of some odour sources. *Analyst*, 106:166–171.

Imachi H., Sekiguchi Y., Kagamata Y., Hanada S., Ohashi A., and Harada H. (2002) *Pelotomaculum thermopropionicum* gen. nov., sp. nov., an anaerobic, thermophilic, syntrophic propionate- oxidizing bacterium. *Int J Syst Evol Microbiol*, 52:1729–1735.

Imachi H., Sekiguchi Y., Kamagata Y., Ohashi A., and Harada H. (2000) Cultivation and in situ detection of a thermophilic bacterium capable of oxidizimg propionate in syntrophic association with hydrogenotrophic methanogens in a thermophilic methanogenic granular sludge. *Appl Environ Microbiol*, 66:3608–3615.

Inanc B., Matsui S., and Ide S. (1999) Propionic acid accumulation in anaerobic digestion of carbohydrates: an investigation on the role of hydrogen gas. *Water Sci Tech*, 40:93–100.

Ishii S., Kosaka T., Hori K., Hotta Y., and Watanabe K. (2005) Coaggregation facilitates interspecies hydrogen transfer between *Pelotomaculum thermopropionicum* and *Methanothermobacter thermoatotrophicus*. *Appl Environ Microbiol*, 71:7838–7845.

Kida K., Morimura S., and Sonoda Y. (1993) Accumulation of propionic acid during anaerobic tratment of distillery wastewater from barley-*Shochu* making. *J Ferment Bioeng*, 75:213–216.

Kim J.Y., Woo S.H., Lee M.W., and Park J.M. (2012) Sequential treatment of PTA wastewater in a two-stage UASB process: focusing on p-toluate degradation and microbial distribution. *Water Res*, 46:2805–2814.

Kivistö A.T. and Karp M.T. (2011) Halophilic fermentative bacteria. *J Biotechnol*, 152:114–124.

Klocke M. (2011) Monitoring microbial communities in biogas plants. In *Biogas Microbiology 2011*, Lepizig, Germany.

Knoblauch C., Sahm K., and Jorgensen B.B. (1999) Psychrophilic sulfate-reducing bacteria isolated from permanently cold Arctic marine sediments: description of *Desufofrigus oceanense* gen. nov., *Desulfofrigus fragile* sp. nov., *Desulfobaba gelida* gen. nov., sp. nov., *Desulfotalea psychrophila* gen. nov., sp. nov. and *Desulfotalea arctica* sp. nov. . *Syst Bacteriol*, 49:1631–1643.

Kolb B. and Ettre I.S. (1997) *Static headspace- gas chromatography. Theory and practice.* . New Jersey: John Wiley-VCH Inc.

Kus F. (1993) Kinetik des anaeroben abbaus von essig- und propionsure in bioreaktoren mit immobilisierten bakterien. Master's thesis, Technische Universität Düsseldorf.

Kus F. and Wiesmann U. (1995) Degradation kinetics of acetate and propionate by immobilized anaerobic mixed cultures. *Water Res*, 29:1437–1443.

Larreta J., Vallejo A., Bilbao U., Alonso A., Arana G., and Zuloaga O. (2006) Experimantal design to optimise the analysis of organic volatiole compounds in cow slurry by headspace solid-phase microextraction-gas chromatography-mass spectrometry. *J Chromatogr A*, 1136:1–9.

Lawrence A.W. and McCarty P.L. (1969) Kinetics of methane fermentation in anaerobic treatment. *J Water Pollut Control Feder*, 41:1–17.

Lebuhn M., Effenberger M., Munk B., Fröschle B., Bauer C., and Gronauer A. (2011) Agricultural biogas production in Germany - from basics to practice. In *Biogas Microbiology 2011*, Lepizig, Germany.

Lens P.N.L., O'flaherty V., Dijkema C., Colleran E., and Stams A.J.M. (1996) Propionate degradation by mesophilic anaerobic sludge: degradation pathways and effects of other volatile fatty acids. *J Ferment Bioeng*, 82:387–391.

Leueders T., Pommerenke B., and Friedrich M.W. (2004) Stable-isotope probing of microorganisms thriving at thermodynamic limits: syntrophic propionate oxidation in flooded soil. *Appl Environ Microbiol*, 70:5778–5786.

Liehr T. (2009) *Fluorescence in situ hybridization (FISH) application guide.* . Berlin Heidelberg: Springer-Verlag.

Lien T., Madsen M., Steen I.H., and Gjerdevik K. (1998) *Desulfobulbus rhabdoformis* sp. nov., a sulfate reducer from a water-oil separation system. *Int J Syst Bacteriol*, 48:469–474.

Lins P., Malin C., Wagner A., and Illmer P. (2010) Reduction of accumulated volatile fatty acids by an acetate-degrading enrichment culture. *FEMS Microbiol Ecol*, 71:469–478.

Liu Y., Balkwill D.L., Aldrich H.C., Drake G.R., and Boone R.D. (1999) Characterization of the anaerobic propionate degrading syntrophs *Smithella propionica* gen. nov., sp. Nov. and *Syntrophobacter wolinii*. *Int J Syst Bacteriol*, 49:545–556.

Liu Y., Xu H.-L., Yang S.-F., and Tay J.-H. (2003) Mechanisms and models for anaerobic granulation in upflow anaerobic sludge blanket reactor. *Water Res*, 37:661–673.

Lowry O.H., Rosenbrough N.J., Farr A.L., and Randall R.J. (1951) Protein measurement with the Folin phenol reagent. *J Biol Chem*, 193:265–275.

Loy A., Lehner A., Lee N., Adamczyk J., Meier H., Ernst J., Schleifer K.-H., and Wagner M. (2002) Oligonucleotide microarray for 16S rRNA gene-based detection of all recognized lineages of sulfate-reducing prokaryotes in the environment. *Appl Environ Microbiol*, 68:5064–5081.

Lozecznk S., Sparling R., Clark S.P., VanGluck J.F., and Oleszkiewicz J.A. (2012) Acetate and propionate impact on the methanogenesis of landfill leachate and the reduction of clogging components. *Biores Technol*, 104:37–43.

Lücker S., Steger D., Kjeldsen K.U., MacGregor B.J., Wagner M., and Loy A. (2007) Improved 16S rRNA-targeted probe set for analysis of sulfate-reducing bacteria by fluorescent *in situ* hybridization. *J Microbiol Methods*, 69:523–528.

Ma J., Carballa M., Van De Caveye P., and Verstraete W. (2009a) Enhanced propionic acid degradation (EPAD) system: Proof for principle and feasibility. *Water Res*, 43:3239–3248. a.

Ma J., Mungoni L., Verstraete W., and Carballa M. (2009b) Maximum removal rate of propionic acid as a sole carbon source in UASB reactors and the importance of the macro- and micro-nutrients stimulation. *Biores Technol*, 100:3477–3482.

Malin C. and Illmer P. (2008) Ability of DNA content and DGGE analysis to reflect the performance condition of an anaerobic biowaste fermenter. *Microbiol Res*, 163:503–511.

Marchaim U. and Krause C. (1993) Propionic to acetic acid ratios in overloaded anaerobic digestion. *Biores Technol*, 43:195–203.

Marta-Alvarez J. (2003) *Biomethanization of the organic fraction of municipal solid wastes*. London. IWA Publishing

Mawson A.J., Earle R.L., and Larsen V.F. (1991) Degradation of acetic and propionic acids in the methane fermentation. *Water Res*, 12:1549–1554.

McCarty P.L. and Smith D.P. (1986) Anaerobic wastewater treatment. *Environ Sci Technol*, 20:1200–1206.

McInerney M.J., Struchtemeyer C.G., Sieber J., Mouttaki H., Stams A.J., Schink B., Rohlin L., and Gunsalus R.P. (2008) Physiology, ecology, phylogeny, and genomics of microorganisms capable of syntrophic metabolism. *Ann NY Acad Sci*, 1125:58–72.

McMahon K., Zheng D., Stams A.J.M., Mackie R.I., and Raskin L. (2004) Microbial population dynamics during start-up and overload conditions of anaerobic digesters treating municipal solid waste and sewage sludge. *Biotechnol Bioeng*, 87:823–833.

Montero B., Garcia-Morales J.L., Sales D., and Solera R. (2010) Evolution of butyric acid and the methanogenic microbial population in a thermophilic dry anaerobic reactor. *Waste*, 30:1790–1797.

Mosche M. and Jordening H.J. (1998) Detection of very low saturation constants in anaerobic digestion: influences of calcium carbonate precipitation and pH. *Appl Microbiol Biotechnol*, 49:793–799.

Müller N., Worm P., Schink B., Stams A.J.M., and Plugge C.M. (2010) Syntrophic butyrate and propionate oxidation processes: from genomes to reaction mechanisms. *Environ Microbiol Rep*, 2:489–499.

Muyzer G. and Stams A.J.M. (2008) The ecology and biotechnology of sulphate reducing bacteria. *Nature Rev Microbiol*, 6:441–454.

Namieśnik J. and Jamrógiewicz Z. (1998) *Fizykochemiczne metody kontroli zanieczyszczeń środowiska*. Warsaw: Wydawnictwo Naukowo Techniczne.

Narihiro T., Terada T., Ohashi A., Kamagata Y., Nakamura K., and Sekiguchi X. (2012) Quantitative detection of previously characterized syntrophic bacteria in anaerobic wastewater treatment systems by sequence-specific rRNA cleavage method. *Water Res*, 46:2167–2175.

Nayono S.E. (2009) *Anaerobic digestion of solid waste for energy production*. PhD thesis, Universität Karlsruhe (TH).

Nielsen H., Daims H., and Lemmer H. (2009) *FISH handbook for biological wastewater treatment: Identification and quantification of microorganisms in activated sludge and biofilms by FISH*. London: IWA Publishing.

Nilsen R.K., Torsvik T., and Lien T. (1996) *Desulfotomaculum thermocisternum* sp.nov., a sulfate reducer isolated from a hot North Sea oil reservoir. *Int J Syst Bacteriol*, 46:397–402.

Oude Elferink S.J.W.H., Boschker H.T.S., and Stams A.J.M. (1998) Identification of sulfate reducers and syntrophobacter sp. in anaerobic granular sludge by fatty-acid biomarkers and 16S rRNA probing. *Geomicrobiol J*, 15:3–17.

Pan L., Adams M., and Pawliszyn J. (1995) Determination of fatty acids using solid-phase microextraction. *Anal Chem*, 67:4396–4403.

Park W., Hyun S.-H., Kim T.-H., and Kim I.S. (2009) Quantitative analysis of the trophic groups with a fluorescence in-situ hybridization (FISH) and the competitive kinetics of a propionate enriched anaerobic culture. *Biotechnol Bioprocess Eng*, 14:523–530.

Peldszus S. (2007) *Handbook of water analysis*. Florida: CRC Press Tylor and Francis Group, 2nd edition.

Percheron G., Bernet N., and Moletta R. (1999) Interactions between methanogenic and nitrate reducing bacteria during the anaerobic digestion of an industrial sulfate rich wastewater. *FEMS Microbiol Ecol*, 29:341–350.

Plugge C.M., Balk M., and Stams A.J.M. (2002) *Desulfotomaculum thermobenzoicum* subsp. *thermosyntrophicum* subsp. nov., a thermophilic, syntrophic, propionate-oxidizing, spore- forming bacterium. *Int J Syst Evol Microbiol*, 52:391–399.

Plugge C.M., Dijkema C., and Stams A.J.M. (1993) Acetyl-CoA cleavage pathway in a syntrophic propionate oxidizing bacterium growing on fumarate in the absence of methanogens. *Microbiol Lett*, 110:71–76.

Pullammanappallil P., Chynoweth D., Lyberatos G., and Svoronos A. (2001) Stable performance of anaerobic digestion in the presence of a high concentration of propionic acid. *Biores Technol*, 78:165–169.

Ravenschlag K., Sahm K., Knoblauch K., Jorgensen B.B., and Amann R. (2000) Community structure, cellular rRNA content and activity of sulfate-reducing bacteria in marine Arctic sediments. *Appl Environ Microbiol*, 66:3592–3602.

Rittmann B., Krajmalnik-Brown R., and Halden R. (2008) Pre-genomic, genomic and post-genomic study of microbial communities involved in bioenergy. *Nature Rev Microbiol*, 6:604–612.

Samain E., Dubourguier H.C., and Albagnac G. (1984) Isolation and characterization of *Desulfobulbus elongatus* sp. nov. from a mesophilic industrial digester. *Syst Appl Microbiol*, 5:391–401.

Sassaki G.L., Souza L.M., Serrato R.V., Cipriani T.R., Gorin P.A.J., and Iacomini M. (2008) Application of acetate derivatives for gas chromatography-mass spectrometry: novel approaches on carbohydrates, lipids and amino acids analysis. *J Chromatogr A*, 1208:215–222.

Sbarciog M., Loccufier M., and Noldus E. (2010) Determination of appropiriate operating strategies for anaerobic digestion systems. *Biochem Engineering J*, 51:180–188.

Schievano A., D'Imporzano G., Malagutti L., Fragali E., Ruboni G., and Adani F. (2010) Evaluating inhibition conditions in high-solids anaerobic digestion of organic fraction of municipal solid waste. *Biores Technol*, 101:5728–5732.

Schink B. (1988) *Biology of anaerobic microorganisms*, chapter Principles and limits of anaerobic degradation: environmental and technical aspects, pp 771–846. Wiley-Liss., New York.

Schink B. (1992) *The Prokaryotes*, chapter Syntrophism among prokaryotes, pp 276–299. Springer Verlag, New York, 3rd edition.

Schink B. (1997) Energetics of syntrophic cooperation in methanogenic degradation. *Microbiol Mol Biol Rev*, 61:262–280.

Schink B. and Stams A. (2006) *The Prokaryotes*, vol. 2, chapter Syntrophism among prokaryotes, pp 309–335. Springer, New Jersey.

Schmidt J.E. and Ahring B.K. (1993) Effects of hydrogen and formate on the degradation of propionate and butyrate in thermophilic granules from an upflow anaerobic sludge blanket reactor. *Appl Environ Microbiol*, 59:2546–2551.

Schmidt J.E. and Ahring B.K. (1995) Interspecies electron transfer during propionate and butyrate degradation in mesophilic, granular sludge. *Appl Environ Microbiol*, 61:2765–2767.

Schmidt J.E. and Ahring B.K. (1999) Immobilization patterns and dynamics ofacetate-utilizing methanogens immobilized in sterile granular sludge in upflow anaerobic sludge blanket reactors. *Appl Environ Microbiol*, 65:1050–1054.

Scholten J.C.M. and Conrad R. (2000) Energetics of syntrophic propionate oxidation in defined batch and chemostat cocultures. *Appl Environ Microbiol*, 66:2934–2942.

Sekiguchi Y., Kamagata Y., Nakamura K., Ohashi A., and Harada H. (1999) Fluorescence *in situ* hybridization using 16S rRNA-targeted oligonucleotides reveals localization of methanogens and selected uncultured bacteria in mesophilic and thermophilic sludge granules. *Appl Environ Microbiol*, 65:1280–1288.

Sekiguchi Y., Kamagata Y., Syutsubo K., Ohashi A., Harada H., and Nakamura K. (1998) Phylogenetic diversity of mesophilic and thermophilic granular sludges determined by 16S rRNA gene analysis. *Environ Microbiol*, 144:2655–2665.

Shi-yi L. and Jian C. (1992) The contribution of interspecies hydrogen transfer to the substrate removal in methanogenesis. *Process Biochem*, 27:285–289.

Shigematsu T., Era S., Mizuno Y., Ninomiya K., Kamegawa Y., Morimura S., and Kida K. (2005) Microbial community of a mesophilic propionate-degrading methanogenic consortium in chemostat cultivation analyzed based on 16S rRNA and acetate kinase genes. *Environ Biotechnol*, 72:401–415.

Shin S.G., Lee S., Lee C., Hwang K., and Hwang S. (2010) Qualitative and quantitative assessment of microbial community in batch anaerobic digestion of secondary sludge. *Biores Technol*, 101:9461–9470.

Sieber J.R., McInerney M.J., and Gunsalus R.P. (2012) Genomic Insights into syntrophy: the paradigm for anaerobic metabolic cooperation. *Annu Rev Microbiol*, 66:429–452.

Stadnitskaia A., Muyzer G., Abbas B., Coolen M.J.L., Hopmans E.C., Baas M., van Weering T.C.E., Ivanov M.K., Poludetkina E., and Sinnighe Damste J.S. (2005) Biomarker and 16S rDNA evidence for anaerobic oxidation of methane and related carbonate precipitation in deep-sea mud volcanoes of the Sorokin Trough, Black Sea. *Mar Geol*, 217:67–96.

Stahl D.A. and Amann R. (1991) *Nucleic acid techniques in bacterial systematics*, chapter Development and application of nucleic acid probes, pp 205–248. John Wiley & Sons Ltd.,, Chichester.

Stams A.J.M. (1994) Metabolic interactions between anaerobic bacteria in methanogenic environments. *Antonie van Leeuwenhoek*, 66:271–294.

Stams A.J.M. and Dong X. (1995) Role of formate and hydrogen in the degradation of propionate and butyrate by defined suspended cocultures of acetogenic and methanogenic bacteria. *Antonie van Leeuwenhoek*, 68:281–284.

Stams A.J.M., Grolle K.C.F., Frijters C.T.M.J., and van Lier J.B. (1992) Enrichment of thermophilic propionate-oxidizing bacteria in syntrophy with metanobacterium thermoautotrophicum or Methanobacerium thermoformicium. *Appl Environ Microbiol*, 58:346–352.

Stams A.J.M., Grothenius J.T.C., and Zehnder A.J.B. (1989) *Recent Advances in Microbial Ecology*, chapter Structure-function relationship in granular sludge, pp 440–445. Japan Scientific Societies Press, Tokyo.

Stams A.J.M., Kremer D.R., Nicolay K., Weenk G., and Hansen A.T. (1984) Pathway of propionate formation in *Desulfobulbus propionicus*. *Arch Microbiol*, 139:167–173.

Stams A.J.M. and Plugge C.M. (2009) Electron transfer in syntrophic communities of anaerobic bacteria and archaea. *Nature Rev Microbiol*, 7:568–577.

Stams A.J.M., van Dijk J.B., Dijkema C., and Plugge C.M. (1993) Growth of syntrophic propionate-oxidizing bacteria with fumarate in the absence of methanogenic bacteria. *App Environ Microbiol*, 59:1114–1119.

Stouthamer A.H. (1979) The search for correlation between theoretical and experimental growth yields. *Int Rev Biochem*, 21:1–47.

Stronach S.M., Rudd T., and Lester J.N. (1986) *Anaerobic digestion processes in wastewater treatment*. Berlin: Springer.

Stubner S. (2004) Quantification of Gram-negative sulphate-reducing bacteria in rice field soil by 16S rRNA gene-targeted real-time PCR. *J Microbiol Methods*, 57:219–230.

Szczepaniak W. (2005) *Metody instrumentalne w analizie chemicznej*. Warszawa: Wydawnictwo naukowe PWN SA.

Talbot G., Topp E., Palin M., and Masse D. (2008) Evaluation of molecular methods used for establishing the interactions and functions of microorganisms in anaerobic bioreactors. *Water Res*, 42:513–537.

Tatara M., Makiguchi T., Ueno Y., Goto M., and Sode K. (2008) Methanogenesis from acetate and propionate by thermophilic down-flow anaerobic packed-bed reactor. *Biores Technol*, 99:4786–4795.

Tchobanoglous G., Theisen H., and Vigil S. (1993) *Integrated solid waste management. Engineering principles and management issues*. McGraw-Hill Inc., Singapore.

Thauer R.K., Jungermann K., and Decker K. (1977) Energy conservation in chemotrophic anaerobic bacteria. *Bacteriol Rev*, 41:100–180.

Thiele J.H., Chartrain M., and Zeikus J.G. (1988) Control of interspecies electron flow during anaerobic digestion: role of floc formation in syntrophic methanogenesis. *Appl Environ Microbiol*, 54:10–19.

Thiele J.H. and Zeikus J.G. (1988) Control of interspecies electron flow during anaerobic digestion: significance of formate transfer during syntrophic methanogenesis in flocs. *Environ Microbiol*, 54:20–29.

Tschech A. and Pfenning N. (1984) Growth yield increase linked to caffeate reduction in *Acetobacterium woodii*. *Arch Microbiol*, 137:163–167.

Van Kuijk B.L.M. and Stams A.J.M. (1995) Sulfate reduction by a syntrophic propionate-oxidizing bacterium. In *Int. Meeting on anaerobic Processes for Bioenergy and Environment*, vol. 5, Copenhagen, Denmark.

Vavilin V.A., Rytow S.V., Pavlostathis S.G., Jokela J., and Rintala J. (2003) A distributed model of solid waste anaerobic digestion: sensitivity analysis. *Water Sci Technol*, 48:147–154.

Wagner M. and Haider S. (2012) New trends in fluorescence in situ hybridization for identification and functional analyses of microbes. *Curr Opin Biotechnol*, 23:96–102.

Wallrabenstein C., Gorny N., Springer N., Ludwig W., and Schink B. (1995a) Pure culture of *Syntrophus buswellii*, definition of its phylogenetic status,and description of *Syntrophus gentianae* sp. nov. *Syst Appl Microbiol*, 18:62–66.

Wallrabenstein C., Hauschild E., and Schink B. (1994) Pure culture and cytological properties of *Syntrophobacter wolinii*. *FEMS Microbiol Lett*, 123:249–254.

Wallrabenstein C., Hauschild E., and Schink B. (1995b) *Syntrophobacter pfennigii* sp. nov., new syntrophically propionate-oxidizing anaerobe growing in pure culture with propionate and sulfate. *Arch Microbiol*, 164:346–352.

Wang L., Zhou Q., and Li F.T. (2006) Avoiding propionic acid accumulation in the anaerobic process for biohydrogen production. *Biomass Bioenergy*, 30:177–182.

Wang Q., Kuninobu M., Ogawa H.I., and Kato Y. (1999) Degradation of volatile fatty acids in highly efficient anaerobic digestion. *Biomass Bioenergy*, 16:407–416.

Wang Y., Zhang Y., Wang J., and Meng L. (2009) Effects of volatile fatty acids concentrations on methane yield and methanogenic bacteria. *Biomass Bioenergy*, 33:848–853.

Widdel F. and Pfennig N. (1982) Studies on dissimilatory sulfate-reducing bacteria that decompose fatty acids II. Incomplete oxidation of propionate by *Desulfobulbus propionicus* gen.nov., sp. nov. . *Arch Microbiol*, 131:360–365.

Wolin M.J. (1975) Interactions between H_2-producing and methane-producing species. In Schlegel H.G., Gottschalk G., and Pfennig N. (eds), *Symposium on microbial production and utilization of gases*, Göttingen, pp 141–150.

Wu W.M., Jain M.K., de Macario E.C., Thiele J.H., and Zeikus J.G. (1992) Microbial composition and characterization of prevalent methanogens and acetogens isolated from syntrophic methanogenic granules. *Appl Microbiol Biotechnol*, 38:282–290.

Zellner G., Busmann A., Rainey F.A., and Diekmann H. (1996) A syntrophic propionate-oxidizing, sulfate reducing bacterium from a fluidized bed reactor. *Syst Appl Microbiol*, 19:414–420.

Zellner G. and Neudörfer F. (1995) Stability and metabolic versatility of propionate-degrading biofilm operating in anaerobic fluidized bed reactor. *J Ferment Bioeng*, 80:389–393.

Zhang S., Cai L., Koziel J.A., Hoff S., Clanton C., Schmidt D., Jacobson L., Parker D., and Heber A. (2010) Field air sampling amd simultaneous chemical and sensory analysis of livestock odorants with sorbent tubes and GC-MS/olfactometry. *Sensors Actuators B*, 146:427–432.

Zhao H., Yang D., Woese C.R., and Bryant M.P. (1990) Assignment of *Clostridium bryantii* to *Syntrophospora bryantii* gen. nov., nov. comb., based on 16S rRNA sequence analysis of its crotonate-grown pure culture. *Int J Syst Bacteriol*, 40:40–44.

Zheng D., Angenent L.T., and Raskin L. (2006) Monitoring granule formation in anaerobic upflow bioreactors using oligonucleotide hybridization probes. *Biotechnol Bioeng*, 94:458–472.

Zhou W., Wu B., She Q., Chi L., and Zhang Z. (2009) Investigation of soluble microbial products in a full-scale UASB reactor running at low organic loading rate. *Biores Technol*, 100:3471–3476.

Zumstein E., Moletta R., and Godon J.J. (2000) Examination of two years of community dynamics in an anaerobic bioreactor using fluorescence polymerase chain reaction (PCR) single-strand conformation polymorphism analysis. *Environ Microbiol*, 2:69–78.